I, Maybot

I, MAYBOT

The Rise and Fall

JOHN CRACE

This edition first published in the UK in 2017
by Guardian Books, Kings Place, 90 York Way, London N1 9GU
and Faber & Faber Ltd, Bloomsbury House,
74–77 Great Russell Street, London WC1B 3DA

Printed in the UK by CPI Group (UK) Ltd, Croydon CR0 4YY

A CIP record for this book is available from the British Library

ISBN 978-1-78335-143-5

6 8 10 9 7 5

For Tom and Debby

Introduction

With the coalition coming to the end of its five-year term of office in May 2015, David Cameron hatched a plan to keep the Conservative Eurosceptics on message during the general election campaign: he would offer them the carrot of a referendum on the UK's membership of the European Union within two years. It was a plan that worked rather too well. Cameron had anticipated the Tories either losing the election or returning to power in another coalition. Either way, the promise of an EU referendum would get quietly shelved.

Instead, the Conservatives were returned with an overall majority and Dave was forced to put up. With the early opinion polls showing a roughly 65–35% split in favour of remaining in the EU, Dave saw no point in delaying the referendum and spent the latter half of 2015 and early 2016 meeting with the 27 heads of the other EU countries in the hope of coming back with a new four-point deal on UK membership that he could present to the country as a marked improvement. The only problem was that the deal looked, to all intents and purposes, identical to the one we already had.

By now, though, Dave was committed and, as there was no backing down, he took to the airwaves to sell a deal whose details seemed a little vague even to him. First up was The Andrew Marr Show, *the BBC's flagship Sunday morning politics show. He got off to a sluggish start. The key phrases his advisers had prepped him with over breakfast – 'important work', 'national security', 'uncertainty' and 'very, very dangerous' – tripped off the tongue as if on time delay. The spirit was willing, but the flesh was weak.*

Marr pressed him on benefits. How could he guarantee that the Department for Work and Pensions would be up to withholding migrants' benefits when it still hadn't managed to get its universal credits system up and running after five years? The only answer to be found in Dave's brain was that with Iain Duncan Smith sacked as its minister, the DWP would run like clockwork, but somewhere in the depths of his hippocampus, Dave knew he wasn't allowed to say that. So he just burbled something meaningless instead.

Slowly, though, Dave's body began to respond and his pulse rose above flatline. Dave remembered that Dave was pumped. He had gone to Brussels and he had bloody well given the EU a kicking. Angela Merkel! Donald Tusk! François Hollande! Whatever the name of the Polish president is! We gave your boys one hell of a beating!

'Can you look me in the eye on sovereignty, prime minister?' Marr asked. No problem. Dave's eyes opened

wider and remained fixed on Marr as he went into a long spiel about how, if we left the EU, our sovereignty would only be an illusion of sovereignty. With Dave we could be in the bits of Europe we wanted to be in and not in the bits we didn't. Paris was fine, but we could give Lille a miss. Stick with Dave and things could only get meta.

On the plus side, Dave knew who his enemies were as the Tory Eurosceptics and UKIP were all known quantities. At least he thought they were. Having apparently promised Dave that he was a committed Remainer, Boris Johnson had a last-minute change of heart while playing tennis with his sister, Rachel. BoJo had become BoGo. This tilted the odds considerably, as Boris – despite his Bullingdon Club background – had always been immensely popular with voters. They liked his style and forgave him misdemeanours that would cost other politicians their careers.

Boris could say what he liked. And frequently did. Demented bureaucrats forcing us to under-power notres vacuum cleaneurs! Why shouldn't we be allowed to crash our camions into ponts if we wanted to, rather than have them restricted to four metres in height? If the EU had its way no one would be allowed to talk Franglais ever again. And that was often pretty much the entire message.

Were there any downsides to leaving the EU? Absolument non. The Frogs and the Krauts would still be gagging for all our clobber and we could keep out any foreigners we didn't like. Un no-brainer. A few people were confused by this. If it had all been that simple why had he apparently agonised for so long over which way to jump? Was it possible he was just in it for himself? Was he just positioning himself to be the next prime minister whatever the result of the referendum? Têtes you win, queues you lose. 'That's an outrageous thing to suggest,' said BoGo, crossing his fingers tightly behind his back. 'This isn't about me.' If true, it would have been the first thing BoGo has come up against that wasn't.

While Nigel Farage and UKIP made the referendum all about immigration, Boris and Michael Gove made it all about money. Lots of it. If we left the EU, we would have an extra £350 million to give to the NHS each week. Even if we didn't. The treasury select committee – one of the few oases of objective facts throughout the referendum campaign – asked Vote Leave to remove the £350 million slogan from the side of its campaign bus on the grounds that it was grossly untrue. Vote Leave smiled politely and said, 'If it's all the same to you, we'll leave it up there as we're not that bothered about how many lies we tell.'

Not that the Remain side were campaigning with any great conviction. Jeremy Corbyn had always been a Eurosceptic when he had been on the backbenches and made a reluctant and underwhelming advocate for remaining in the EU now he was Labour leader. Several prominent Conservatives (including Theresa May – lest we forget) were notionally on the Remain side, but managed to avoid saying almost anything in three months, leaving Dave and Chancellor George Osborne to do all the heavy lifting. Their tactic was not to sell the benefits of remaining in the EU so much as terrifying everyone with the consequences of leaving.

'I'm not about Project Fear,' Dave declared firmly. 'I'm about Project Fact'.

The difference between Project Fear and Project Fact wasn't immediately obvious. A greater emphasis on frightening facts, rather than factual fears.

Of course everything would still be OK if Britain did decide to leave the EU but no one should be under any illusions that it would be a Great Leap in the Dark, a world of continual night where the country would be left at the bottom of every global pile going. Trade deals? We might get something negotiated within a decade, but there were no guarantees. That kind of thing. Those kinds of facts.

With the Remain camp running a dismal, joyless campaign and Vote Leave promising the Earth, the polls began to narrow significantly. Two weeks before polling day, the first poll came out predicting a Leave victory. The Remainers were visibly panicked but even on 23 June, the day of the referendum, no one on either side really thought anything but a narrow vote to stay in the EU was on the cards. The electorate had other ideas. Whether because voters really did want 'to take back control' or because they just fancied sticking two fingers up to the establishment, the UK voted by 52% to 48% to leave the EU. And so a tumultuous year began . . .

Over to you, says puffy-eyed Cameron as the Brexit vultures circle

24 JUNE 2016

Shortly after 6 a.m. a van pulled up outside Downing Street. With still no sign of David Cameron, who had been expected to make a statement minutes earlier, the hordes of photographers gathered outside the prime minister's front door snapped the newspaper delivery man instead. Something to do. This was history and no one wanted to miss a moment.

There was still no sign of the prime minister nearly an hour later when someone opened the door of No. 10 to let Larry the Downing Street cat out for his morning stroll. The photographers got their cameras out again. Larry sat on the porch for five minutes, wondering if he was about to be the fall guy in a dead cat bounce. Surprised to find himself still alive, he exited stage right.

Another half an hour passed and Larry reappeared. The front door opened and he went back inside. Still no sign of Dave. It was becoming startlingly clear that No. 10 was in crisis. The prime minister knew he should have made a statement long ago, but he still didn't really have a clue what he was going to say. What could he say? He'd gambled the future of the country for an internal party squabble and he'd lost.

As sterling dropped another few cents, a French broadcaster rehearsed her lines. *'David Cameron est fini,'* she said. 'David Cameron is finished.' It was apparently going to be a dual-language broadcast. On the other side of the gates at the far end of Downing Street, an organ grinder played an off-tempo version of 'Land of Hope and Glory' while passing cars honked their horns in approval. The Brexit vultures were closing in.

At 7.40 a.m. Lord Feldman, the Conservative party chairman, knocked on the front door of No. 10. He was kept waiting outside considerably longer than Larry. The new world order was making itself felt. 'Have you got anything to say, Lord Feldman?' a reporter shouted. He hadn't. No one else seemed to have anything to say, so why should he?

Still no Dave. At 8 a.m. the financial markets opened and £100 billion was wiped off their value within minutes. So much for the prime minister calming City nerves. Shortly before 8.30, Dave's favourite oak lectern was carried out into the street, and moments later he and his wife Sam walked out. Both looked puffy-eyed. It had been a long night, and the day was going to be even longer.

'Good morning everyone,' he said, grasping the lectern with both hands. 'The will of the British people is an instruction that must be delivered. Across the world people have been watching the choice that Britain has made. I would reassure those markets and investors that Britain's economy is fundamentally strong.'

His body language was anything but reassuring, and neither was his implication that the British people had come to the wrong decision. He wasn't the right man to lead the negotiations for this country's exit from the EU, he continued, so he would be standing down as prime minister before the Conservative party conference in October.

It was said with dignity as well as edge. Scotland was already seeking a second independence referendum to keep the country in the EU, Northern Ireland might do the same, Spain was making claims on Gibraltar and Britain faced years of economic uncertainty. If Boris Johnson and Michael Gove were so sure they could sort out this mess, they were welcome to have a go. His job had stopped being fun and he'd had enough.

'I love this country, and I feel honoured to have served it,' he said, his voice beginning to crack. Only a huge effort of will got him to the end. 'And I will do everything I can in future to help this great country succeed. Thank you very much.' Dave and Sam instinctively moved to kiss each other. At the last second they caught one another's eye and thought better of it. To touch would only lead to more tears. They deserved some dignity.

Over at Vote Leave headquarters, Boris and Gove were looking equally stunned. Neither had either expected to win or Cameron to resign, and what had started out as a bit of a game had become horribly real. Sombre faces were the order of the day. Boris began by paying tribute to Dave – 'He's been a great prime minister and his only fault was to have the job I wanted' – before trying to appeal to the young people who had resoundingly rejected him. Having spent the entire campaign ignoring the young, he couldn't help but sound unconvincing. Sincerity has never been Boris's strongest suit.

Gove's shock felt rather more real. He looked like a man who had just come down off a bad trip to find he had murdered one of his closest friends. But even he couldn't avoid hypocrisy. After openly rubbishing each and every expert for weeks, he tried to reassure everyone that everything was going to be OK because Brexit would be entrusted to great minds. Both declined to take any questions from the media. Which was just as well, because right now they didn't have a single answer.

* * *

As half the country celebrated the referendum result and the other half reeled in shock, it quickly dawned on everyone that the political classes were totally unprepared for what came next. Some were calling for Article 50 to be triggered immediately, others were insisting on a delay while a few seemed to be in denial that the referendum had ever taken place.

Into this vacuum stepped . . . another vacuum, as all the key players in the referendum campaign seemed to go missing. Though you might have thought that Boris and Gove, the two main architects of the Leave campaign, would have been only too keen to share their cunning plans with a divided nation, the merry pranksters had other things on their mind. Lunch. Cricket. Lunch. That sort of thing. Why was everyone so keen on hearing about a plan? Couldn't they understand that Brexit was never about having a plan? It was about taking back control of not having a plan.

Nor were the Remainers in any better shape. Some professed amazement that anyone had actually taken them seriously when they had said they supported Britain staying in the EU, while the best that David Cameron could come up with in the Commons was that there was almost unanimous agreement within the Cabinet to appoint some experts to try to come up with a vague plan about how to make an even vaguer plan to think about how Britain was going to leave the EU. Indeed, it rather looked as if the main priority for the Conservative party wasn't Brexit at all. It was choosing a new leader.

* * *

Boris's career undone by a Poundland Lord and Lady MacGove

30 JUNE 2016

Even the most sophisticated of wife-swapping parties can end in tears. The Conservative annual fundraiser at the Hurlingham Club had started so well. Car keys had been flung into a solid silver bowl and Leavers and Remainers were happily getting off with one another. 'I know I called you a duplicitous, lying moron during the referendum campaign,' they cooed to one another, 'but deep down, darling, you know I've always fancied you.' The champagne flowed, the beds bounced, the sheets squelched.

Then in walked Sarah Vine, Michael Gove's wife. Sarah wasn't at all happy. She didn't mind that her husband was copping off with someone else – hell, he'd been cheating on everyone in the Conservative party for as long as she could remember. What really upset her was that Mikey was being so poorly rewarded for his infidelity. Her husband was selling his body far too cheaply. If he was going to sleep with Boris he should at least be guaranteed one of the great offices of state. The treasury, for instance. So what if Mikey was hopeless at maths and couldn't even be trusted with the shopping?

Sarah donned her Marigolds, strode into a back bedroom and consciously uncoupled Mikey. Back in their west London house, she gave Mikey a strong talking to.

'Look, Mikey,' she said. 'I didn't support Vote Leave just for you to get shafted by Boris. You might have thought the referendum was all about the country but it wasn't. It was all about me. So get your finger out.'

On Thursday morning at 9 a.m., Mikey duly obliged by releasing an email saying although he had promised the country countless of times that he definitely wouldn't be standing for leadership of the Conservative party he had now decided to stand for leadership of the Conservative party because Sarah had another *Daily Mail* column to write and fancied being in the government. Mikey tried to warn Boris in advance, but unfortunately his phone was out of charge. He was sure Boris would understand, though.

Over at St Ermin's Hotel in Westminster, Boris was busy preparing to launch his own leadership campaign. The bunting was out and so were his supporters: Nadhim Zahawi, Crispin Blunt, David Davis and Jesse Norman. The first sign of the impending shit storm was the arrival of Zac Goldsmith, a man who has never knowingly backed a winning side. The second was a minder asking everyone to wait outside the room. 'There's tea and coffee over there,' she said, helpfully.

'Is the announcement still going ahead?' I asked.

'There's tea and coffee over there.'

Twenty minutes later than planned, we were all let in and Boris was greeted with a standing ovation by his acolytes. But this wasn't the Boris they had come to know and love. This was a Boris whose gags were falling flat.

A Boris whose heart just wasn't in it. A Boris who could quote a speech from Julius Caesar seemingly oblivious to the irony of treachery. A Boris who was used to being the stabber, not the stabbee. A Boris whose career had been undone by a Poundland Lord and Lady MacGove.

Boris couldn't resist a bit of drama, delaying the announcement that he wouldn't after all be standing for the Conservative party leadership right till the end, but that was just about the only sign of life he gave. The deflation was near total. Boris had only ever come out for Leave to get the top job and now his ambition was dust. Nadine Dorries was in tears by the end; so it wasn't all bad.

Within minutes of Boris standing down, his closest supporters were saying they would now be backing anyone but Gove. Unbelievably, the two men who had campaigned so hard and often so unpleasantly together during the referendum were now engaged in open warfare. Brexit had never been wholly about Brexit. It had been a disguise for a Tory party leadership contest. Thanks for nothing, boys. I hope it was worth it.

Theresa May had looked much more chipper at her own leadership launch in the establishment setting of the library of the Royal United Services Institute on Whitehall. Having spent most of the referendum campaign supporting Remain by saying nothing, she was understandably a bit croaky – her vocal cords are still recovering from lack of use – when presenting herself as the woman who could unite the country and take Britain out of the EU.

'I'm a straightforward kind of politician,' she said. 'I don't indulge in gossip, I just get on with the job.' Well, not that much. Earlier in the week, she had lunch with the *Daily Mail* editor Paul Dacre, but I am sure they were just chatting about the football. Had she known Boris was going to step down, she'd have toned down some of the barbs – 'The only deal he has ever done with the Germans is to buy three creaky water cannons' – and directed them against Mikey. But there would be time enough for that. Out of one *ménage à trois* with Boris and Mikey and into another with Mikey and Sarah. Things were about to get even messier.

* * *

In the subsequent leadership race, Stephen Crabb and Liam Fox were knocked out in the first round. Crabb because no one knew who he was and Fox because everyone knew exactly who he was. Gove predictably fell at the next hurdle, leaving a straight fight between Andrea Leadsom and Theresa May. Leadsom had mysteriously risen without trace. In the first outing of her campaign she had seemed to imply that her main strength was that she, unlike May, was a mother, and in a subsequent event that was billed as a 'major speech on the economy' she had barely mentioned the economy.

Her 'Rally4Leadsom', consisting of Tim Loughton, Theresa Villiers and a handful of bemused stragglers walking a couple of hundred yards to Westminster shouting,

'Who do we want?'

'Andrea Leadsom!'

'When do we want her?'

'Sometime quite soon!'

was the highlight of the Tory leadership campaign – for the political sketch-writers, if not for Leadsom. Within days, she had given an interview to The Times *in which she had repeated her insistence that being a mother made her better qualified, and this time the ensuing outcry forced her to drop out.*

The Tory leadership race had turned into a race for the bottom and was over before it had even really got going, with Theresa May becoming prime minister by default. Just as in the referendum campaign, where she had remained largely mute, May's success owed as much to her silence as anything else. She had only made one short speech, in which she had said disobliging things about Boris – and even this had been entirely unnecessary given Boris's inglorious exit from proceedings immediately afterwards – and no one had really even thought to question her less than impressive credentials as home secretary.

* * *

Boris? Michael? Andrea? Theresa rules the roost after manic Monday

11 JULY 2016

Just another manic Monday. Little more than a day after receiving a text from Andrea Leadsom saying: 'Soz I sed u wld b rubbish leedr cos u is not a mum,' Theresa May walked into a meeting of the 1922 Committee in Portcullis House to be anointed as the next prime minister. Ten minutes later she left to a standing ovation from Conservative MPs trying to outcompete one another in expressions of undying devotion. Boris who? Michael who? Andrea who?

If Theresa looked a little nonplussed when she appeared with her husband and inner circle of loyal MPs outside St Stephen's Gate, she wasn't the only one. Even by recent standards this was all a bit quick. Only the Conservatives can combine the brutality of a Stalinist purge with the low comedy of a *Carry On* film. It had trusted the country to reach the right decision in the referendum campaign and it wasn't going to make the same mistake again by giving the untamed fringes of the Tory party a say.

'Honoured and humbled,' she mumbled. 'Brexit means Brexit.' Though not necessarily, if the man by her side, Chris Grayling, was to become minister for Brexit. Grayling has yet to find a ministerial job he can't do slowly and badly. Having said the bare minimum,

Theresa scarpered off home to wonder how a day that had started off with her launching her leadership campaign in Birmingham had ended with her landing the top job. Seldom had so much been achieved in British politics by saying and doing so little.

The first sign that Westminster had accidentally overdosed on speed yet again was when the net curtains twitched at Andrea Leadsom's leadership campaign headquarters in Westminster shortly after midday. Moments later Steve Baker, Owen Paterson, Iain Duncan Smith and Tim Loughton trooped out the front door to form a *Dad's Army*-esque Praetorian guard on the doorstep. Then came Andrea.

The wider Andrea's smile became, the more furious IDS looked. Quantity theory in action. 'I have a statement which is mine which I wish to read out,' Andrea smiled. IDS lowered his eyes. If anyone said a word out of place, the pavement was going to get it. Forget the velvet glove of compassion. There was more than one career here that was about to be shot down in flames.

'It has only just come to my attention I have the support of just 25% of Conservative MPs,' she continued, forcing the words through the fixed smile, 'and that, in these uncertain times, the country doesn't need a nine-week leadership campaign.

'I've also taken a look at the people around me and decided most of them are an electoral liability. So I have decided to withdraw my name from the contest and let

Theresa May be prime minister. Sorry to have made such a nuisance of myself. I'm now going to lie down in a dark room.'

Andrea declined to take any questions, so we never did get to find out how it had taken her four days to work out that most of her own MPs thought she was far too hopeless to be leader, when everyone else had done the sums in a matter of seconds. Percentages can't have been her strong point when she was working in the Barclays call centre.

It was left to Andrea's cracked troops to pick up the pieces. Most chose to jump ship at the earliest opportunity. 'Andrea has been absolutely brilliant but I've always secretly thought Theresa was the right man, sorry woman, for the job, party must unite blah blah and if you're recording this then I am definitely interested in any jobs that might be going.'

Only Loughton remained faithful to the Belle Dame *sans* so much as a *merci*. 'There have been dark forces at work,' he muttered. He meant journalists accurately reporting answers freely offered, but might just as well have been referring to the Tory party itself.

'Another Brexiter leaves the scene of the crime,' yelled a passerby who had just happened to catch the tail end of Leadsom's speech. He had a point. One by one, the prime architects of the Vote Leave campaign had managed to stab one another in the back, front and sides, and now the last one standing had thrown herself onto the funeral pyre.

Back at No. 10, David Cameron was on the phone

to his therapist trying to deal with his self-destructive issues when he heard that Theresa was going to be moving in a great deal earlier than anticipated. 'Bugger it,' he yelled. It just wasn't fair. Now he wouldn't get to fly in his brand-new Dave Force One plane to Africa. Now he'd miss his last G20. Now he'd have to find somewhere to rent as he'd given his tenants notice to leave in September. The way the day was going, George would forget to bring back a suitcase full of dollars from New York.

'I'm off on Wednesday afternoon,' he announced grumpily to the single camera parked outside the Downing Street door. 'Good luck to the lot of you.' Dave took a couple of deep breaths, trying to calm himself down as he marched back inside. It was no good. He was still furious. Perhaps humming might help. The theme to *The West Wing* somehow felt appropriate. Then for the removals. And cut.

Choosing a cabinet might be fun after all, thought Theresa

13 JULY 2016

It hadn't been the easiest of meetings with the Queen. After observing that they both had husbands called Philip, the conversation had rather died. Eventually, the Queen had broken the silence and asked, 'And what do you do?' Theresa May had been nonplussed by that. She didn't do anything. Doing nothing had been the only quality she had needed. One by one her opponents had offed themselves in ever more ridiculous circumstances until she was the last one standing. 'I suppose I must be the prime minister,' she said, kneeling before the Queen.

The car journey back from the palace hadn't exactly been a bundle of laughs, either, once she'd read David Cameron's leaving speech. What was it he had said again? 'I'm leaving Britain a stronger country.' Was he mad? It was precisely because the country was in such a mess that Dave had been bundled out at short notice. Still, a little graciousness wouldn't go amiss. Social niceties weren't her strong suit, but she could probably rustle something up.

As her grey government Jag parked up in Downing Street for the first time, she walked briskly across the street towards the prime ministerial lectern with her husband in tow a couple of paces behind. 'David Cameron has done a brilliant job in uniting the country in fury at the clueless way he has handled the referendum,' she began.

Theresa thought about adding something about the government's budget promises being left in tatters, but decided against it. This wasn't the time for small talk. She reached into her back pocket for the speech she had given at her Birmingham leadership launch. No one would notice the recycling, as Andrea Leadsom's self-immolation two days previously had meant no one had bothered to report it.

'If you're black . . . If you're white working-class . . . If you're a woman . . . Life can be a struggle,' she said. 'My government will not be for the privileged few.' Up in north London, Ed Miliband choked on his tea. This was exactly the speech he had intended to give if he had won the general election the year before.

Beyond the Downing Street gates, a group of protesters chanted: 'What do we want? Brexit. When do we want Article 50? Now', but Theresa ignored them. She knew full well that fudging an exit from the EU was at the top of her in-tray. There was no point in making life any more difficult for herself by making promises she might not be able to keep.

Speech over, she retreated for the obligatory doorstep photograph with her husband. 'Try to smile,' Philip said. 'Why should I?' she replied through pursed lips. The snappers weren't satisfied. They wanted the money shot. 'Give her a kiss,' they yelled. 'Give her a kiss.' Theresa faced them down. She was the prime minister, not some performing seal. Who did they think she was? David Cameron?

Once inside No. 10, Theresa checked her phone. The inevitable obsequious tweet about how sodding marvellous she was from Matt Hancock, who had never yet found a bum in which he didn't want to place his nose. Tough Matt. No big job for you. Then she called Philip Hammond; he was dull enough to be given the treasury.

Next up, Boris. People always said she didn't have a sense of humour; well she'd prove them wrong. She'd always intended to ditch the public school boys, but everyone would enjoy Boris getting a hospital pass of foreign secretary. Let's see how he got on with all the foreigners he'd managed to insult over the years. Choosing a cabinet was more fun than she had imagined.

The jostling for position had started at prime minister's questions, with all the Tories taking their seats much earlier than usual. Killing two birds with one stone: waving Dave off and making a good impression in front of Theresa. Some were just a bit too quick off the mark. Poor old Greg Clark, the secretary for communities and local government, who had got in early doors, was kicked off the front bench to make way for more deserving causes. And more desperate ones.

Liam Fox was leaving nothing to chance; having been the first to be kicked out of the Tory leadership race on a 'Brexit means going to war with the Hun' ticket, he has been clinging to Theresa like a limpet ever since. Fearful he might get lost in the crowd of backbenchers behind her, Fox had positioned himself directly in her eyeline in the overspill gallery. On his head was a neon sign, flashing 'See Me, Feel Me, Touch Me, Gissa Job'. 'Chill, Liam,' she mouthed. 'Needy isn't a good look.'

The rest were left to take their chances as Theresa made her entrance into the chamber. No one wanted to be the first person to stop cheering for fear of appearing disloyal, so the applause went on for far longer than was strictly necessary. The Law of Inverse Disloyalty. George Osborne used the few minutes he had unlimited access to Theresa's right ear to make more bantz than in the last six years. 'Hi, it's me,' he said. 'I've always loved working with you.' Theresa ignored him. He didn't know it yet, but he was toast.

Theresa slumped back in her seat. This was about as good as it was going to get. Brexit was bound to end in tears. Her career was bound to end in tears. Politics was like that. But she did have one advantage. At least she had one thing going for her. There was no opposition. At PMQs the entire Labour party had been playing Pokémon Go on their mobiles, desperately hunting down a leader. For now the trickiest bastards were all behind her. Even if they weren't yet with her.

Talk to the hand, Leadsom: Theresa May's perfect first day

14 JULY 2016

Theresa May bounced into her Westminster office. She could have just sacked the dead wood over the phone like most prime ministers have done, but why deny herself the pleasure of doing it in person? First in the queue outside her door was Michael Gove.

'Hello, you treacherous little shit,' she said, evenly.

'I've never liked you. Let alone trusted you. You're fired.'

'Please don't,' Mikey whimpered. 'Sarah will kill me if I come back with nothing. I'll do anything. Junior minister in transport . . .'

'Next.'

Next was Nicky Morgan. 'Can you give me one good reason why I shouldn't get rid of you?' Theresa snapped. Nicky's mouth opened and closed without saying anything. Same as it always did.

'Next.'

In came Oliver Letwin. 'You're sacked.'

'Really? I didn't even know I had a job.' Oliver had never been the most worldly of politicians.

'Brutal reshuffle,' shouted the hacks gathered in Downing Street as Theresa walked back into No. 10. Theresa grinned for the cameras. Yes, it had been and she'd enjoyed every second. She had waited years to settle some of these scores. Now for the equally fun bit of dispensing favours that could be cashed in later.

Liz Truss was first through the door. Liz had gone down a storm at the previous year's party conference, venting her fury at cheese. Who better to put in charge of justice? 'Freedom for the Wensleydale Four,' Liz shouted cheerfully on her way out.

Then came a nervous-looking Jeremy Hunt. Understandably. Theresa had already briefed the BBC that he was going to be sacked.

'I know I've been totally useless and I've messed up

with the junior doctors,' Jeremy pleaded. 'But I can do better. I promise.'

'Lucky for you, no one else wanted your poisoned chalice,' said Theresa, narrowing her eyes to mere slits. 'But you're on a final warning. One more cock-up and you're toast.'

'Oh thank you, thank you.'

Jeremy couldn't help punching the air for the photographers as he walked out of No. 10. 'I haven't been sacked,' he yelled. Everyone was just as astonished as he appeared to be.

Time for a breather and some lunch. Theresa switched on the news to find that, in his first address to Foreign Office staff, Boris Johnson had promised to re-colonise Africa and pose naked as Mr November for President Putin's 2017 calendar. Maybe it hadn't been quite such a good appointment after all.

Patrick McLoughlin came and went fairly quickly. He didn't look nearly as pleased to be offered the post of chairman of the Tory party and chancellor of the Duchy of Lancaster as Theresa had expected. 'But you'll be able to appoint a few vicars and magistrates, Patrick,' she said to his departing back.

More of a problem was Stephen Crabb, who had spent the previous 90 minutes chained to a radiator, after being informed he was being sent back to the Welsh office from the Department for Work and Pensions. 'I'm a Crabb,' he pointed out, reasonably. 'I'm used to going sideways, but I'm buggered if I'll go backwards.'

'The thing is this, Stephen,' Theresa said. 'Your sexting wasn't a good look. It's Wales or nothing.'

'It's nothing.'

'As you wish. Consider yourself a hermit Crabb.'

Chris Grayling also appeared underwhelmed to be put in charge of roadworks on the A303 and making sure HS2 was indefinitely delayed. 'But I was your right-hand man throughout your leadership campaign,' he wept.

'But Chris, you always said you weren't doing it in the hope of any reward.'

'I didn't mean it, though.'

'Tough. It's the best I can do. They don't call you Failing Grayling for nothing.'

Theresa's phone rang. It was one of her special advisers, reminding her she had forgotten to appoint someone in charge of the Department for International Development. She groaned. There was just so much to remember. 'Do we have anyone who has actively campaigned to abolish the department?' she asked.

'Yup. Priti Patel.'

'Great. Give her the job.'

By now it was getting late in the day, but Theresa had got her second wind. Besides, she had been saving the best till last.

'Bring me the head of Andrea Garcia,' she commanded.

Andrea Leadsom duly trooped through the door.

'You've been a right pain in the neck during the

referendum campaign and since,' Theresa observed. Andrea tried to get in a word to explain how she hadn't meant any of it, but Theresa just showed her the palm of her hand.

'It's been brought to my attention that your extensive knowledge of farming has led you to state that hill sheep smallholdings should be converted into butterfly sanctuaries,' Theresa said. 'As a result of this, I have decided it's only right to put you in charge of the environment, farming and rural affairs. Now go and explain to the farmers how they're going to be worse off without their EU subsidies.'

'But I hate the countryside . . .'

'GO.'

The perfect end to the perfect first day.

Theresa could have reinvented herself as anyone – but she came as Maggie

20 JULY 2016

The Thatch is back. For her first prime minister's questions Theresa May could have been anyone. She could have been Sensitive Theresa, Caring Theresa, Funny Theresa. Any Theresa she cared to reinvent herself as. Instead, she said: 'Tonight, Matthew, I'm going to be Maggie.' Why be your own woman when you can be the one whom large sections of the Tory party have never fallen out of love with?

Close your eyes and the years could have been rolled back to the early 1980s. An uncompromising, graceless and brittle figure at the dispatch box and a horde of semi-priapic backbenchers braying. The Tories might have a far better record at appointing female leaders than Labour but their male MPs still leave a lot to be desired.

The similarities with the mid-80s didn't end with a female prime minister. The Labour benches are in as much disarray now as they were then, with a leader who fails to inspire any confidence in his own MPs. Jeremy Corbyn's arrival in the chamber was greeted with barely a flicker of interest by even his frontbench. Corbyn tried to pretend this was all normal as he prepared for his first confrontation with May but he couldn't conceal his humiliation.

Corbyn started promisingly by holding the prime minister to account for her speech outside Downing Street the week before, in which she had laid out her social justice programme. Yet somehow, despite holding all the aces – after all, May had been home secretary for six years in a government under which inequality had significantly worsened – Corbyn failed to land any killer punches.

For a brief moment, it had looked as if he might score heavily with a reference to Boris Johnson's track record of casual racism and open insults to every country on the planet but he missed the opportunity to make the point stick. Instead, May was allowed to get away with just ignoring her foreign secretary. For now. Boris is going to come in for this kind of flak wherever he goes and his position as a serious diplomatic negotiator must be untenable.

Beginning to realise Corbyn had nothing to offer that could hurt her, May began to channel her inner, hard-core Thatch. 'You call it austerity,' she growled, in a voice chillingly reminiscent of her predecessor. 'I prefer to call it living within our means.' Here was the grocer's daughter made flesh. The ordinary housewife – albeit one, like Maggie, who was also conveniently married to an extremely supportive millionaire – who looked after the shillings and pence on the nation's weekly shopping bill. At the first sighting of Iron Lady 2.0, the more incontinent Tory backbenches had their first premature ejaculation.

Thrilled to have negotiated her first sticky patch, Theresa went for 110% Maggie. When Corbyn made the schoolboy error of bringing up job insecurity, May reached for her one pre-scripted gag. 'I suspect that many members on the opposition benches might be familiar with an unscrupulous boss – a boss who does not listen to his workers, a boss who requires some of his workers to double their workload and maybe even a boss who exploits the rules to further his own career,' she said.

Each word was dragged out with the comic timing of someone failing an audition for Britain Hasn't Got Talent but the Tories greeted it with hysterical laughter. A mixture of creepy unctuousness and childhood regression to Mummy. 'More,' they cried. Theresa obliged. Lowering her voice to a register even Maggie might have struggled with and adopting an exaggerated, pantomime gurn, she added: 'Remind you of anyone?' Behind her, several other backbenchers rushed out to get some wet wipes.

Thrilled the Cameron shackles were now off and that the happy nasty party days were back in vogue, the Conservative Stuart Andrew celebrated by making a gay joke. 'Growing up on a council estate, I found it tough coming out – as a Conservative.' Boom, boom, Stuart. The old ones are the old ones. On the front bench, Boris looked rather annoyed that someone else was getting the laughs. Perhaps the Commons wanted to hear his one about the piccaninnies and Nurse Ratchet?

The nursery was becoming rowdier and rowdier. For May, this was all getting a bit too cosy and familiar. Time to remind people who was boss. She looked around for the weakest person in the Commons. After Tim Farron had offered her a gracious welcome, May responded by humiliating him. 'My party is much bigger than yours,' she sneered. It was classless, graceless and unnecessary but it still provoked roars of approval. Blessed are the meek, for they will be roundly trashed.

* * *

After the seismic shocks of the EU referendum and the Tory leadership election, the summer recess came as a welcome break for politicians from all parties. Theresa May chose to go walking in Switzerland for her holidays. Those hoping some mountain air might refresh the Supreme Leader's rhetoric would be sorely disappointed, however. At the first cabinet after recess she said, 'We must continue to be very clear that Brexit means Brexit, that we're going to make a success of it. That means there's no second referendum, no attempts to sort of stay in the EU by the back door, that we're actually going to deliver on this.' This lack of clarity might not have done her any harm in the polls – An ICM/Guardian poll gave the Conservatives a 14-point lead over the opposition – but it wasn't likely to impress any of the other world leaders at the G20 summit in China she was due to meet in a few days' time.

* * *

So Brexit means Brexit means Brexit. Is that it?

5 SEPTEMBER 2016

The six-week holiday may have gone some way to concentrating the mind, but it has done little to clarify the thinking. Brexit remains as gnomic now as it did back in July. 'The reason I've been saying Brexit means Brexit is precisely because it means it does,' said Theresa May, pioneering a new branch of illogical positivism during a rather tetchy press conference at the G20 summit in China.

What Brexit had appeared to mean at the G20 was the prime minister getting shunted to the back row of the leaders' group photo, being briefed against by the Americans and the Japanese and being left to big up the fact that Mexico, Australia and Singapore have expressed a vague interest in doing trade deals with the UK. It's a start, I suppose. If not the one that May would have been hoping for.

Nor was there any real enlightenment on the meaning of Brexit to be found in the Commons as Brexit minister David Davis gave his first lack-of-progress report. This was Davis's first outing on the government front bench for more than 19 years and he came to the house flanked by Boris Johnson and Liam Fox as his security blanket. This unexpected show of unity was quickly explained; none of them have yet actually done anything to fall out over. Give it time. Quite a long time, to judge by Davis's statement.

'Britain voted overwhelmingly to leave the EU,' he began. In Brexitworld a 52–48 vote is a total landslide. 'So Brexit means Brexit means Britain leaving the EU.'

It wasn't long before the Labour benches started laughing and shouting, 'Waffle, waffle.' Davis took this as an instruction rather than a criticism. 'We will be creating beacons and roundtables of organisations,' he waffled on. 'There will be challenges but these are opportunities and everything will basically be fine once we've got round to thinking about it with the brightest and best minds in Whitehall, though obviously there can be no room for complacency.'

'Is that it?' interrupted the SNP's Pete Wishart. Davis nodded. That was about it, though he was more than happy to repeat himself for another five minutes or so before concluding that he would be returning to parliament at regular intervals to give updates on everything that wasn't happening.

When Emily Thornberry was first appointed shadow minister for Brexit alongside her day job as shadow foreign secretary it looked as if the reason she had been made to double up was because Jeremy Corbyn hadn't been able to find anyone else willing to do it. Now the duplication looks more like an act of genius. Why bother to have a separate shadow minister for a department that wasn't likely to be doing very much for the foreseeable future?

'So far all we've learnt about Brexit is that the government is not going to introduce a points-based immigration system or give £350 million per week to the NHS,' she observed. 'Both of which were two of the key Vote Leave promises in the referendum campaign. The government has gone from gross negligence to rank incompetence. You're making this up as you're going along.'

Davis took this as a compliment. A sign that he was really getting to grips with the job and that progress was being made. Even if only by a process of elimination. 'We're definitely not going to have a points-based system because that is what the prime minister said yesterday,' he declared. 'What we are going to have is a results-based system that might be even tougher.' There again, it might not. It was precisely to sort out these kinds of details that he would be consulting roundtables and beacons.

Thereafter, the house divided on predictably partisan lines. Those on the Remain side wanted to get to grips with the nitty gritty of what access to the single market Britain would get, how EU laws would be repealed and whether Britain would remain signed up to Europol. Those on the Leave side thought such things were minor niggles and what really mattered was sticking two fingers up to the Frogs and the Hun and returning sovereignty to parliament. Though not to the extent of giving parliament a vote on the details – should any ever emerge – of the Brexit negotiations reached as it might vote against it.

'Is that it?' several more MPs enquired.

'Is that it?' Davis echoed. He's been on the back benches for so long he hasn't quite appreciated he's now supposed to be answering the questions not asking them.

Not quite. There was just time for newspaper columnist Michael Gove to declare that everything was going far better than 'the soi-disant experts with oeuf on their face' had predicted and begging the minister never to consult anyone who might know what they were talking about. So far, that's the one promise Davis has been able to keep.

Brexit means never having to say you're sorry (or anything at all)

7 SEPTEMBER 2016

Careless talk costs lives. With her advisers having belatedly realised that saying 'Brexit means Brexit' was providing the country with far too much information about what Brexit actually means, Theresa May devoted much of prime minister's questions and her subsequent statement on the G20 summit in China to damage limitation. She has already had to slap down David Davis for making up policy on the hoof in the Commons and her other two liabilities, Boris Johnson and Liam Fox, have yet to open their mouths.

It used to be the case that most people stopped listening to PMQs once the two leaders had finished going head to head. On Wednesday, it was the moment when many chose to tune in. With Brexit uppermost in everyone's minds and the government front benches struggling even to maintain the 'Brexit means Brexit' line, Jeremy Corbyn asked the prime minister about the housing crisis. It was almost as if he was making a point of ignoring Owen Smith's email instructing him to lead with Britain's relationship with Europe.

May couldn't believe her luck and clumsily shoehorned the gags she had prepared for harder questions into her non-answers. Her escape was only temporary, though, as she was forced on to the back foot by the SNP's Angus Robertson, who wanted to know whether Britain would remain part of the single market. She glanced at her notes. 'REMEMBER NOT TO SAY BREXIT MEANS BREXIT'. Written in capitals.

'Brexit means . . .' she began, before pausing. What did it mean if it didn't mean Brexit? No. She couldn't allow herself to think that way. She must try to stifle her natural tendency towards honesty and transparency. She started again. 'It would not be right to give a running commentary on our Brexit. I know this is exactly the opposite of what the minister for Brexit promised you all on Monday, but that only goes to show how much progress we have made with our Brexit negotiations over the course of two days.

'The relationship with Europe that we will be getting is a very special one that I can't tell you about right now because I haven't got a clue what it will look like. But I can promise you that it won't be a Norwegian relationship because we are not Norwegians.

'It will be a very special British relationship, which will be ours and ours alone and, once I have spent the next two years failing to get what I want, I will tell you what I have reluctantly settled for. Trust me on this, though. You're wanting it. You're loving it. You're getting it.'

If May appeared somewhat taken aback when she eventually realised that many MPs were openly laughing at her, she looked abject when Tory backbencher Bernard Jenkin leapt to her rescue. 'I feel more confident now about the future of the UK than at any other time in my life,' he announced. Jenkin is one of those politicians with the unerring knack of being wrong about almost everything. So when he's feeling good, it's time for the rest of us to panic.

There was no let-up when the prime minister moved on to her G20 statement. 'It would not be right for me to give a running commentary on our Brexit negotiations,' she said, more than happy to repeat herself. While it wasn't right to give a running commentary on her Brexit negotiations, it was perfectly in order to give a running commentary on why she wouldn't be giving a running commentary. 'And not giving a running commentary was the process I took into the G20.

'But let me say this. I am delighted to say that Mexico and Singapore are quite keen to do a trade deal with us at some unspecified time in the future and that when the Australians say they are going to put the UK at the back of the queue behind the EU, what they really mean is we are at the top of their thoughts and prayers. And by the way, I did whisper something about steel dumping when I was in the toilet so nobody can say I wasn't tough with the Chinese.'

Most MPs had hoped May might have had a little more to show for her jaunt to China and several tried to tease out a few more details about what Brexit deals were in the offing. 'I understand that you don't want to give away any sensitive information,' said Conservative Anna Soubry. 'But could you at least tell us some of the principles that will underlie the negotiations?'

'No.'

Labour's Yvette Cooper tried another tack. 'Without giving away any sensitive information, could you give us an idea of the values that will inform the negotiations?'

'No.'

No principles. No values. No progress. No clue.

* * *

It hadn't just been the Tories that had been having a leadership contest. Having only just got himself on to the ballot paper at the last minute the year before, thanks to a few Labour MPs thinking it would look good to have a token lefty in the mix, Jeremy Corbyn found himself facing a second contest inside 12 months. With Labour consistently lagging a long way behind the Tories in the polls, Owen Smith had put himself forward as the moderate candidate in a new leadership challenge. The result, which was announced at the Labour party conference in Liverpool, was the same as before: though the country didn't appear to think too highly of Corbyn, the Labour membership (dramatically increased since the reduction of the joining fee to £3) was overwhelmingly behind him.

'I would like to thank Owen Smith,' Corbyn had said

in his acceptance speech. 'We've had some interesting and good-natured debates.' This wasn't exactly the way anyone else remembered the debates, in which both men had repeatedly traded threats and insults, calling one another 50 shades of useless – but the victor always gets to rewrite history.

As Corbyn moved off the internal divisions of the party and on to fighting the Tories, the atmosphere relaxed a little. Sensing he now had the crowd behind him, Corbyn went on to give one of his most convincing speeches, talking about the need to fight against grammar schools, tackle child poverty and hold the government to account over its Brexit negotiations. If he had sounded quite so passionate at key moments earlier in the year, he might have avoided going through a second leadership election.

'We can win the next general election,' he concluded, unintentionally splitting the party he had just spent the last 10 minutes trying to unite. His supporters cheered. They believed they could and would win, if only the British people let slip their collective false consciousness. The moderates believed the opinion polls and reckoned they were doomed to defeat while Corbyn remained their leader.

The Tories also believed they were bomb-proof so long as Corbyn was Labour leader, and his re-election took the edge off what could have been a tricky Conservative party conference for Theresa May. Beneath the surface, the old tensions over Europe remained. Some wanted a total split, others wanted to remain in the single market and the customs union. Though few Tory politicians were prepared to say that Brexit was a bad idea in public, there was intense disagreement in private about the terms on which Britain should leave the EU.

In her end-of-conference speech, the prime minister tried to unite both sides. Not entirely successfully. As 'Start Me Up' pumped through the sound system, Theresa May danced out of the shadows hellbent on making some grown men cry. This was her Year Zero. Everything that had happened in the past six years had been nothing to do with her. She had hated every last minute. She had been home secretary only in name. She had been a prisoner in her own department. Any pro-EU sentiments she might once have voiced had been implanted in her head by metropolitan liberal elite aliens. And now a change was gonna come. Sam Cooke would have felt far queasier about having his song lyrics hijacked than the Rolling Stones.

'I'd like to pay thanks to the man who made the party change,' she said. It sounded very much as if she was going to give a shout out to Nigel Farage, but the man she had in mind was David Cameron. The former prime minister had changed things a lot in the past six years. For the worse. So it was up to her, Theresa, to change everything all over again. Dave was a stain on the country and his legacy needed to be erased. Theresa had just made her first grown man cry.

A change was gonna come. The country had spoken and she was listening. She had made no great efforts to pay attention to the feelings of those who had been left behind during the referendum campaign, but she was now. Brexit must mean Brexit, and she could guarantee she would get the best deal for Britain. She didn't say how because she wasn't going to give a running commentary, but the rest of the world would inevitably bow to the might of British sovereignty. The Labour party might be heading back to the 1970s, but this was an unashamed retreat to the 1870s.

A change was gonna come. It was just as well no one had bothered to enquire if the change would be for the worse.

* * *

The PM's Brexit confusion is contagious

12 OCTOBER 2016

Theresa May was confused. She didn't appear to have heard her home secretary telling the Conservative party conference that foreign workers would be named and shamed. 'That never happened,' she insisted at prime minister's questions, 'which is why I went out of my way earlier in the week to say that when Amber Rudd had said this she actually meant the complete opposite.' Doublethink used to be a prerogative of the left.

The prime minister was also confused about whether parliament should be allowed to debate the terms on which Britain would negotiate its exit from the EU. 'When I said the government wasn't prepared to debate Brexit,' she again insisted, 'what I really meant was that it hadn't occurred to me parliament would be interested in why the value of the pound was falling further and further by the day.'

Most of all, though, Theresa was confused by the Labour party. For months now she had been used to it being a bit of a rabble and squabbling among itself. Now it was showing signs of getting its act together. Jeremy Corbyn was asking trickier and more focused questions and the backbenchers were more interested in getting under Tory skins than their leader's. No one had ever warned her that the opposition might actually oppose and she couldn't handle it. She felt brittle and unsure; even she wasn't convinced by her answers.

Grumpy David Davis was also confused. And even grumpier than usual. He'd been told the government wouldn't be giving a running commentary on its Brexit negotiations and here he was being dragged back to the Commons for the second time in three days to explain why he didn't really have a clue about what he was supposed to be doing and was making it up as he went along. He hadn't voted for Britain to leave the EU only for parliament to hold him to account. That's not what he called 'taking back control'.

Labour's shadow Brexit minister, Keir Starmer, tried to make things easy for Grumpy. Scrutinising the government's Brexit negotiations was not about trying to renege on the result of the referendum; it was about trying to negotiate the best possible outcome for leaving the EU. Hadn't Grumpy tabled a 10 minute rule bill back in 1999 which questioned the right of the executive to ratify treaty changes without a vote in parliament?

'La la la,' grumped Grumpy. 'I'm not listening.'

'Allow me to put it this way, then,' Starmer continued, helpfully. If Grumpy was going to act like an idiot, he would treat him like one. 'The referendum was a mandate to leave the EU. It wasn't a mandate on the terms on which we would leave. You can't build a consensus around a position that you've refused to disclose.'

'Another day, another outing,' said Grumpy when he finally got to have his say. 'My mandate is my mandate. It is the biggest mandate ever. Take back control. I am not a robot. Clunk. Clunk. Whirr.' Grumpy didn't bother to check his notes. He was still on auto-pilot from last Monday. 'There's no need for the government to do anything it doesn't want to. We've done quite enough voting. My mandate is my mandate. Control back take. Clunk. Whirr. Plop.'

'Actually,' Labour's Jack Dromey interrupted. 'A junior minister, George Eustice, has just gone on the radio to say that the government will be putting forward a green paper for parliament to debate. So that rather suggests we will get a vote.'

'Mandate. Back control take. There will be no vote. Ever. Over my rapidly dying body. Clunk. Clunk. Whirr.' Grumpy's system was suffering severe overload from a government making up policy on the hoof behind his back. Grumpy made his excuses and made a quick dash for the exit.

The reason for Grumpy's systems failure soon became even clearer. It wasn't just the opposition that was learning how to oppose. It was also the Tory backbenchers who were beginning to find their voice. 'My duty to represent my constituents transcends that of my duty to the party,' said Dominic Grieve. The sentiment was echoed by Anna Soubry, Nicky Morgan, Claire Perry, Chris Philp, Ken Clarke, Maria Miller and Alastair Burt.

Back at No. 10, Theresa May could see her majority slipping away in front of her eyes. There must be no vote prior to Article 50 being triggered after all. She picked up the phone and told Eustice that when he had talked about putting a Brexit green paper before parliament he'd really been talking about a completely different green paper. Eustice scratched his head. The prime minister's confusion was contagious.

* * *

At the European Council summit in October, Theresa May was increasingly marginalised by other EU leaders who were more interested in discussing Russian aggression in Syria and ongoing trade negotiations with Canada than listening to what Britain had to say about leaving the EU. Brexit was a British problem and until Britain actually set out clear plans for what it wanted to achieve – the only plan on offer at this point was to have no plan – and triggered Article 50, the 27 other EU member states weren't that bothered in having a conversation. To make their feelings plain, the EU made Theresa May wait until the small hours before allowing her just five minutes to talk about Brexit.

* * *

Talk to the hand, Theresa, because the EU aren't listening

21 OCTOBER 2016

It had been a long, long, Thursday evening at the European Council summit for Theresa May. First she had had to listen to loads of boring speeches about things she wasn't very interested in and then she had got stuck next to a French bloke who'd insisted on talking to her in French. What was wrong with the EU these days?

Back in David Cameron's time, everyone used to speak in English.

Eventually her frustration had got the better of her and she had turned to the Frenchman to ask why her speech was only number 15 on the agenda.

'Parce qu'il n'y a pas 16 items *sur* the agenda,' the Frenchman had said, while swigging another tumbler of merlot.

The time passed slowly. Ten o'clock came and went. Eleven o'clock came and went. Midnight came and went. Just after 1 a.m., a steward tapped her on the shoulder to let her know there was a spare five minutes if she had anything she wanted to get off her chest while the few remaining people still awake finished their coffee.

'I'd like to talk to you all about Brexit,' Theresa had begun.

'Parlez à la main,' shouted a lone Belgian, before falling off his chair.

Theresa continued, determined not to be distracted. 'I'm here to tell you that Brexit means Brexit and that the UK remains committed to getting whatever deal with the EU we can manage to negotiate once we've got some sort of a clue what it is we really want. *Merci, danke.*'

Silence. Two people staggered off to bed without saying a word; the rest remained asleep in their chairs. The *épaule* had never seemed so *froide*.

'I think that went quite well on the whole,' Theresa said later to one of her advisers.

'Er, yes . . .' the adviser replied, guardedly. 'I suppose you were at least invited to this summit. We only got to hear about the last one in Bratislava after it had finished.'

Things didn't really pick up that much the following morning. On her way into the meeting, she had heard the European Commission president, Jean-Claude Juncker, whisper he wasn't interested in doing Britain any favours and the frosty reception she had got in the room suggested that was a sentiment shared by other EU leaders.

'Can we talk about Brexit?' she begged.

'*Non.*'

'But I need to be able to tell people back home something. Can I just say we've agreed to start preliminary trade deals?'

'*Nous* will say *rien* until you invoke Article 50.'

'But I won't know what I want unless we have some discussions before I trigger Article 50.'

'*Parlez en français,*' sniggered Michel Barnier, the EU's chief Brexit negotiator.

'*Mon français n'est pas très bon.*'

'*C'est meilleur que votre position de négociation!*'

'He is right,' said Juncker. 'Besides, *nous* want to talk about Russia, not *vous.*'

The meeting ended with Theresa sulking and saying nothing as the Lithuanian president sent her email links to YouTube footage of Russian warships sailing up the English channel.

'It's time for your press conference,' her adviser

reminded her, shortly after 1.30 a.m. 'Don't forget to sound really upbeat and tell everyone this summit has been a huge personal success.'

'This summit has been a huge personal success, and Britain remains a confident, outward-looking and enthusiastic member of Europe,' said a hollow-eyed, flatlining Theresa, sounding diffident and introverted. 'We talked a bit about Russia and immigration and I am sure we can make a success of Brexit so long as people stop making it difficult for me. Now, does anyone have any questions?'

Theresa looked up, hoping that just this once no one did. No such luck. Did she really expect 27 countries to listen to us when we're leaving? Wasn't the EU out to embarrass us and make things difficult for us?

'Um . . . er . . . people really are listening to us,' she said. 'They've just a funny way of showing it. Now is there anyone from the overseas media here?'

A hand shot up. Theresa fell on it gratefully, relieved to be able to show the whole world that at least one other country was listening to her. The hand turned out to belong to another UK journalist who was usually ignored and was trying to blag a question. There really was no one else out there listening to her after all. Theresa had never felt quite so alone.

* * *

On November 3rd, in an action brought by Remain campaigner Gina Miller, the High Court ruled the government was not able to invoke Article 50 by royal prerogative and that parliament must be allowed a substantive vote on the matter. The decision was not well received in government.

'We want to bring back control of our laws to the UK,' David Davis, the Brexit secretary had growled in parliament. Except for those ones which we didn't trust British judges to adjudicate on in the way we would have liked them to. Sovereignty had its limits apparently. 'The government came to the conclusion that it was perfectly OK to invoke Article 50 using prerogative powers,' he continued. Kindness prevented him from commenting on whether Theresa May had made a misjudgement of her own in appointing Jeremy Wright – a man described by former Tory MP Stephen Phillips as 'a third-rate conveyancing lawyer' – as attorney general. 'The court came to a different view and we are disappointed by that.'

Disappointment was the least of it. Fury, outrage and humiliation were more like it. But the increasingly grumpy Davis was in no mood to back down. Having repeatedly told MPs that the result of the referendum must stand and there should be no going back, Davis sounded awfully like someone who was having a major sulk about not getting his way and was going to carry on fighting until he did. 'We're going to the supreme court, who we believe will tell us we are being proper and lawful,' he insisted. And if the Supreme Court let him down, he'd go to the Supreme Supreme Court. And if that failed he'd just appoint some judges of his own. Job done.

The government weren't the only ones to be incandescent about the high court ruling. Many of the right-wing, pro-Brexit newspapers had branded the judiciary as a disgrace. The Daily Mail *had branded the judges 'Enemies of the People'. Anyone who tried to stop the government from doing exactly what it wanted was trying to thwart the will of the people and was therefore an enemy of the state.*

As far as both the government and the press were concerned there could be no question of the government revealing its negotiating strategy to parliament – not least because it doesn't have one – as it would undermine its already ropey position. The idea that the government had already given away a key bargaining position by declaring it would invoke Article 50 before the end of March didn't seem to have occurred to Grumpy.

MPs from both sides of the house invited the government to put forward a resolution to invoke Article 50, in order to allow the house to prove its concerns on this issue were over parliamentary sovereignty and not to delay Britain's exit from the EU. 'No,' snapped Davis. 'We're going to go to the supreme court and that's that.' Giving parliament any say in Brexit would be the thin end of the wedge. Give MPs an inch and they would take a mile.

Theresa May managed to avoid most of the fallout from the government's High Court defeat by being out of the country on a trade mission to India in a bid to prove that Britain had a global future outside the EU. It was while on this trip that her Maybot tendencies first came to the fore . . .

* * *

Theresa struggles to take back control – from her own Maybot

8 NOVEMBER 2016

About halfway through a rather soporific appearance before the public administration and constitutional affairs select committee, the former head of the civil service, Gus O'Donnell, thought it worth reminding everyone that civil servants were more like humans than robots. Which could be just as well, as the prime minister is increasingly acting like someone who is more robot than human. Sometime between July, when she looked like the safest pair of hands amid a sea of idiots, and now, Theresa May's brain appears to have been hacked. Ask her a sensible question and you're now guaranteed a senseless answer.

'Have you made any plans for a Brexit transitional

deal?' inquired a Sky News reporter, at the end of the prime minister's near pointless jolly to India.

Whirr. Clunk. Clang. The Maybot's eyes rotated into life. 'I'm focusing on delivering Article 50,' she replied, unable to prevent herself from answering an entirely different question.

'Will you be able to deliver on the £350 million that was promised to the NHS?' the reporter persevered.

'When the people. Whirr. Voted in the referendum. Clunk. They wanted. Clang. A number of different things,' said the Maybot, struggling with her circuit board.

'Was the referendum dishonest?'

Inside the Maybot, the last shards of the real Theresa were fighting to get out. She was not a number. Especially not 350 million. She was a person in her own right. She did still have a mind of her own. Then the malware took over again.

'Whirr. The referendum took. Clunk. Place. I'm focusing . . .' She wasn't. She really wasn't.

'You weren't part of the Vote Leave campaign, you weren't prime minister at the time of the referendum and you have no mandate,' observed the reporter sharply.

'I'm. Whirr. Determined . . .'

'Stephen Phillips, the MP who resigned last week, said that the Conservative party is becoming more like UKIP. How do you feel about that?'

'I'm. Whirr. Determined,' the Maybot clunked.

'You're determined to be what?'

'I'm. Whirr. Determined. To be. Clunk. Determined to focus on the. Clang. Things that the British public determined . . .'

At this point the Sky reporter cut his losses and left. There was no point in trying to deal with a severe Maybot malfunction.

With the Maybot temporarily on idle, Theresa frantically hammered at the control-alt-delete keys to crash herself, in a last-ditch attempt to return to her factory settings.

'Please ask me about my holiday in India,' she begged.

'Er, no,' said a BBC reporter. 'The Institute of Fiscal Studies is forecasting a £25 billion slowdown. Is that a price worth paying for greater controls of immigration?'

'I'm. Whirr. Determined,' the Maybot laughed, thrilled to have survived the reboot. 'Brexit offers a. Clunk. World of opportunities. I'm determined to be here in India determinedly delivering. Clang. On a determined global Britain through some determined trade deals. Whirr . . .'

'Is an economic slowdown a Brexit price worth paying?' the reporter repeated, generously giving the prime minister the benefit of the doubt that she had not heard the question properly first time round.

'Do you want to see my snaps?' the Maybot whirred. 'There's a great one of me in the hotel lobby with Geoffrey Boycott. Such a sweet man. I've always been a huge fan of his. Who is he again?'

'Thank you, prime minister . . .'

'India is a lovely place. Whirr. And we've been determined to do some. Clunk. Good deals that are not worth the determined paper they are. Clang. Written on as nothing can be determined. Clunk. Before we determine how determined we are to be in a determined customs union . . .'

'What about the slowdown?'

'I'm determined to be. Whirr. Determined . . .'

Theresa knew she was determined. But what about? Slowly it came back to her. Whirr. She was determined to take back control. And she would start by taking back control of her own brain. The Maybot laughed. Some hope.

Hammond warned against Brexit and no one listened. Now it's payback

23 NOVEMBER 2016

'The economy is strong and resilient,' Philip Hammond began. Lurch's face then cracked into a half smile. He'd been only kidding. The economy was actually in a complete mess and he couldn't have been happier. He'd warned his colleagues of the dangers of Brexit and no one had bothered to listen. So now he was going to spell out the consequences to them and they'd just have to sit there and suck it up. The 10 cabinet members sitting alongside him all looked pretty pleased about it too; then they had also all voted to remain in the EU. Coincidence? Hardly. This autumn statement was to be a day of reckoning for the Brexiteers.

First in line was Boris Johnson. 'I suspect I will be no more adept at pulling rabbits from hats than my successor as foreign secretary has been at retrieving balls from the back of scrums,' he sniggered. Lurch knew it was just gratuitous sadism to openly mock Boris's hapless effort to become prime minister, but he was having too good a time to stop himself. Besides, Johnson had managed to annoy just about everyone in the cabinet over the past few months, so giving him a kicking now counts as a team-building exercise in the Tory party.

Lurch then went on to list a litany of failure. Sterling depreciation. Growth slower than expected. Everything the last chancellor had done binned. The budget deficit up to £120 billion. The cost of Brexit an extra £60 billion. Tax receipts lower. Even the lazy French and Italians were more productive than us. 'Members of the house may be interested to know . . .' he said. But they weren't. At least not those on his own benches. They just wanted him to sit down and shut up as soon as possible. This occasion was meant to be an opportunity for the chancellor to boast how brilliant he had been. Not an excuse to admit that everything had gone badly wrong.

Having got most of the really bad news out the way, Lurch moved on to the merely bad news. Old infrastructure schemes that had already been announced got re-announced. The minimum wage was increased by less than the last chancellor had promised. People on welfare would still probably die, the only saving grace being that it might take them a little bit longer to do so.

'I have deliberately avoided giving a long list of projects,' Lurch declared, apparently unaware he had just done so. Quite a lot escapes him.

Half an hour into his speech, most of the house was nodding off. Lurch is one of the few politicians dull enough to make a death spiral sound boring, and vain enough not to notice. After ploughing through a particularly turgid passage in total silence, Lurch thought to interrupt himself. 'That bit was complicated,' he said. 'But it was actually good news.' Theresa gave him an embarrassed nudge. Lurch's empathy skills have often left a little to be desired and he'd mistaken sleep for incomprehension.

By now the few Tory backbenchers who were still alive were getting desperate. They wanted something to cheer. Anything. So when Lurch announced he was going to hand over £7.5 million to save Wentworth Woodhouse, an old pile near Rotherham, they became ecstatic. Getting over-excited about the survival of a stately home wasn't really what they had had in mind, but it was the only orgasm on offer.

Too bad the 'just about managings' wouldn't have enough spare cash to go to visit the place when it was finally done up. There again, they probably wouldn't want to anyway. 'It is said to be the inspiration for

Pemberley in Jane Austen's *Pride and Prejudice*,' Lurch had said. Only it wasn't. Lurch had just gone and saved the wrong palace.

Lurch lumbered slowly on, enjoying the looks of horror on his own side of the house. It was all going even better than he had planned. Most chancellors choose not to leak all their good news stories in advance and save one or two for the speech itself. But Lurch had out-thought everyone. For him the bad news was the good news. The country had voted for Brexit and the country could pay for it. For the next 10 years or longer.

'I do have one last cunning plan,' Lurch said as he came to the final page. Some Tories perked up. Maybe there was to be salvation after all. 'My cunning plan is to rename the autumn statement as the autumn budget and the spring budget as the spring statement.' Genius. Not with a bang but a whimper.

* * *

With most EU countries not really in the mood to talk to Britain until after Article 50 had been triggered – and even then only if they really must – Theresa May had found herself short of countries willing to indulge her desire to shoot the breeze about foreign policy. So when Poland indicated it was willing to have a bilateral meeting, May was only too keen to roll out the red carpet. Schmoozing a right-wing, xenophobic government might not have been the best of looks when Britain was trying to reposition itself as open and friendly to Europe, but beggars can't be choosers.

'We've had an excellent and historic first summit,' said Theresa at her most Maybotic, frantically racking her brains for anything memorable that had been discussed. After saying she was sorry for all the attacks on Poles in the UK since the EU referendum, the conversation had rather dried up.

The Polish PM, Beata Szydło, had looked on impassively as the Maybot ran through her highlights package of the day's events. She recalled it all rather differently. 'Great Britain doesn't have summits with countries like Poland very often,' she observed. And she was looking forward to many more in the coming months. Starting in Warsaw next year. The Maybot looked startled. Had she really agreed to that? The Polish interpreter whispered into her earpiece, assuring her that she had.

Still, at least meeting the Polish PM had taken her mind off the fact that the government was fighting what was clearly a losing battle in the Supreme Court to overturn the High Court ruling on Article 50. If the first day of the hearing hadn't gone badly enough for the government barrister, James Eadie, the second got off to a shocker.

'I think you've just given two diametrically opposed answers to the same question in the last five minutes,' observed Lord Sumption. As Eadie, aka the Treasury

Devil, had only been back on his feet for less than 20 minutes, this wasn't the best of starts.

'We'll have to look back through the transcripts and see which one we agree with then,' Lord Carnwath added, not altogether helpfully.

'I see,' said Lord Neuberger, trying to be kind. 'We had better let you proceed with your argument.'

'I will try not to give two inconsistent answers in the next five minutes,' Eadie said dolefully. He'd never wanted this appeal and just going through the same points that the divisional court had dismissed last time out was doing nothing for his self-esteem. Trying to make the best of a bad job wasn't his usual style.

He began to fumble and lose his way. When he reached the point of double taxation in his submissions, he just decided to give up the unequal. The justices might understand the law but he didn't and he'd only end up giving more wrong answers. 'Because of time,' he said, 'I'll pass on this.' It wasn't quite the slam dunk finale he'd been hoping for.

Lord Pannick, Gina Miller's barrister, only had to open with the observation that 'If the government is right, the 1972 European Communities Act has a lesser status than the Dangerous Dogs Act,' and the case was all but decided.

Things weren't looking any better for the government on any other front. In an appearance before the treasury select committee, Robert Chote, chairman of the

Office for Budget Responsibility, had been questioned by Conservative Jacob Rees-Mogg on his gloomy economic forecasts. How could the OBR be so certain about the levels of uncertainty?

Chote had raised an eyebrow. He had expected to be grilled on the numbers rather than metaphysics and ontology. 'The thing about uncertainty is that it's uncertain,' he said certainly. It was now Rees-Mogg's turn to look puzzled. Chote tried to help him out. It was like this: though he couldn't be quite as certain about the levels of uncertainty as the Bank of England, he was still certain enough about the uncertainty to be confident in the uncertainty. Or to put it another way, uncertainty + uncertainty = certainty.

'We've had to make some assumptions based on government policy even though we and the government know that some policies are never going to be implemented,' said Chote. 'On other matters we asked the government to explain its policies but they didn't seem to have any.'

Nor had David Davis's appearance before the Brexit Select Committee gone much better for the government. David had admitted he couldn't say when the government would have a plan, other than it definitely wouldn't be within the next month as he had 57 sectoral analyses to complete. Some of which were barely under way. Nor could he promise a white paper, nor how many pages the plan would be. It all depended on the font size. 'We just want everything to run smoothly,' he said, hoping that platitudes might be mistaken for thoughtfulness.

'Do you worry about going over a cliff edge?' enquired committee chairman, Hilary Benn.

Davis closed his eyes. He wasn't really sure if he was meant to be that bothered about going over a cliff edge or not. Obviously it wouldn't be a great idea to rush headlong off the cliff but if everyone was to line up in an orderly fashion and then jump off the cliff, surely that couldn't be too bad?

And yet despite all this, the Tories still held a commanding double-figure lead over Labour in the opinion polls. Something not even Theresa May's toe-curling embarrassment at the European Council meeting, in which she was ignored by almost everyone, could dent.

* * *

Theresa May feels the love from her cabinet after unhappy Eurotrip

19 DECEMBER 2016

As away days go, Theresa May's trip to the European Council last week was right down there. The other EU leaders were either ignoring her or laughing in her face.

Instead of getting an invitation to the evening's dinner, the prime minister found herself picking at a cold Unhappy Meal on the Eurostar on the way back to London. Even the Maybot has feelings and it had taken intensive work with her therapist to persuade her to come to the Commons to give a statement on her abject failure.

'Try to reframe the experience,' her shrink had said. 'I know it felt like your first day in the school playground when nobody spoke to you, but there's no need for such a primal regression. It wasn't that everyone else thought you were a total loser, it's just that they were too shy to talk to someone with your charisma. And if it's any consolation, your cuffs did look fantastic after you spent 10 minutes anxiously fiddling with them.'

The Maybot had only been slightly reassured by her therapist's intervention and was still in an extremely delicate state as the clock ticked round to 3.30 p.m. on Monday afternoon. So to make sure she did not back out, all the senior members of her cabinet – Boris Johnson, Philip Hammond, Amber Rudd, Michael Fallon and David Davis – were drafted in to sit next to her. Seldom has so much moral support made a prime minister appear quite so vulnerable.

'The main focus of the EU Council was about how we could all work together,' the Maybot began. Which wasn't necessarily the way the other 27 EU leaders had remembered it, but she was fairly sure that none of them were in London to contradict her. If they had not been bothered to talk to her when they were in the same room, there was little chance of them following her back to London.

NATO. Syria. Holidays in Cyprus. The Maybot scratched her head. She was sure there was something else she had mentioned in the five minutes she had been given while half the room had nipped out for a comfort break, but she couldn't think for the life of her what. Then it came to her: Brexit. 'I reassured them all we were looking forward to a smooth and mature Brexit.' She didn't care to add that the Polish prime minister had joked about whether this smooth 'hot tub' Brexit was the same as a red-white-and-blue Brexit.

Jeremy Corbyn was not too bothered by the imaginary conversations the Maybot had had about NATO and was keen to press her further on Brexit. How had she managed to become so isolated in Europe? Why was her government in such a shambolic mess? Why did one cabinet minister keep promising one thing only for the others to promise something else? Could she promise Britain wouldn't be liable for a £50 billion bar bill on leaving the EU? And when would she be presenting her Brexit plan to the parliament?

The Maybot dabbed her eyes. Isolation had always been one of her key sobbing trigger words in therapy. 'I'm – we're – not isolated,' she said, hastily correcting herself. 'We may be leaving the group but everyone basically adores us and wants to carry on being friends.' BFFs. What she had always craved. She managed to forget to mention the £50 billion. Just as well, probably.

With the opening exchanges over, the Maybot began to relax a little, as all the Brexiteer backbenchers who had been press-ganged into turning up expressed their undying admiration for her genius in choosing not to have a negotiating position for leaving the EU.

Iain Duncan Smith declared that anyone who wanted to know what the hell was going on was being unpatriotic. Peter Lilley was insistent that every day we stayed in the EU was another day when £250 million wouldn't be going into the NHS. Boris was about to correct him that the real figure was £350 million before remembering neither figure was accurate.

May's mood lifted slightly. The session hadn't been as traumatic as she had feared. She may have been taken apart by the opposition benches but she had felt some lurve from her own side. It wasn't real lurve, she knew that. But when you're desperate, any lurve will do.

Theresa May stumbles on a question of thought

20 DECEMBER 2016

There may have been things the prime minister wanted to do less on the last day of parliamentary business before the Christmas recess than appear before the liaison committee, but none immediately came to mind.

Being interrogated on Brexit by the chairs of all the select committees is no one's idea of fun. Especially when you don't have any of the answers.

Andrew Tyrie, the committee chair, got things under way with a few well-aimed questions on the timing of Brexit and the possibility of extending the negotiation period.

'Our intention is . . .' said the Maybot, before answering an entirely different question. It's always so much easier to answer the questions you've thought up yourself, rather than the ones you've been asked.

'I'm trying to get some clarity,' sighed an exasperated Tyrie.

And the Maybot was trying not to provide any. The Maybot is nothing if not pre-programmed. It's just the country's bad luck that she's been pre-programmed to say nothing. Even her words of empty reassurance only manage to inspire a feeling of panic.

Hilary Benn, one of Westminster's kinder and more patient souls, failed to make much headway. Would parliament get to scrutinise the government's plan before she triggered Article 50? Maybe yes, maybe no.

'What would you consider a reasonable period of time?' Benn asked, reasonably. The Maybot shrugged. Depends whether you call a quick glance of a yellow Post-it note scrutiny. And what you call a plan.

'The EU parliament will get a vote on the Brexit negotiations,' Benn continued, doggedly. 'Why can't you guarantee that the UK parliament will also get a vote?'

This produced the same non-answers as before. For someone who has been coded in binary, the Maybot finds it surprisingly difficult to give yes and no answers.

'Do I take it the government will have a standstill arrangement?' Tyrie interrupted.

'I wouldn't say standstill,' said the Maybot.

'Others would,' Tyrie observed drily. 'Is an adjustment period a priority?' The Maybot scratched her head. The computer said no. The computer said yes. The computer said: 'I want my mummy.'

She fought hard to explain her feelings about the EU. It wasn't them that had changed, it was her. The divorce negotiations weren't about keeping the bits of the relationship she liked, they were about creating an entirely different relationship. Albeit one that looked pretty much like the old one but with a bit more casual sex thrown in.

The session turned noticeably tetchy when Yvette Cooper, the home affairs select committee chair, questioned the prime minister on immigration. Cooper and the Maybot have previous unfinished business and there's not much love lost between the two. Was immigration going to be a key part of any Brexit plan? If it turned out that abandoning immigration targets was in the best interests of the country, would she do so?

'You can't look at it like that,' the Maybot snapped. 'The two things aren't linked.'

'That's odd,' Cooper replied. 'Because it was you that first linked them.'

The Maybot fiddled with her pen angrily, trying to make Cooper disappear by the power of thought. No luck. She closed her eyes and tried harder. Still no luck. Why couldn't Cooper understand that many people had only voted to leave the EU because they weren't that keen on foreigners? There was only one thing for it.

'Waffle, waffle,' the Maybot waffled.

'If the net migration targets supersede any other Brexit agreements,' Cooper said, 'can you tell us which people you don't want to come to the UK?'

'Waffle, waffle,' the Maybot waffled.

'You haven't answered any of my questions.'

For the first time in an hour, the Maybot looked relieved. At least she had got one thing right. Though not for long. Sensing weakness, Tyrie moved in for the kill.

'There's just a couple of things I would like to clarify,' he said. 'Will the UK parliament be kept at least as well informed as other EU parliaments?'

'Um,' muttered the Maybot.

'Is that a yes?'

'Um?'

'I'm hearing a no.' Tyrie moved on to transitional arrangements. The chancellor had said thoughtful politicians were in favour of them. Would the Maybot classify herself as a thoughtful politician?

She had to think very hard about this. Was she thoughtful? Like so many things to do with Brexit, it was hard to say. 'When I said I was against transitional arrangements,' she stammered, 'I didn't mean to imply I was against implementational arrangements that are near enough the same thing and might all be ripped up anyway if the other EU countries don't like them and then we'll all be buggered unless we have some transitional arrangements that I definitely don't want.'

The Maybot had just proved herself to be the queen of dialectics. By trying to appear thoughtful, she had achieved the opposite.

* * *

Christmas gave everyone two weeks' breathing space. Both Theresa May and Jeremy Corbyn used the period to deliver almost identical messages to the country: 2016 had been a tumultuous year in politics. The people of Britain had shown they were unhappy with the way Westminster worked; they felt let down by both politicians and the system. It was time to find a new way of doing politics. To deliver a Brexit that worked for everyone. Most ordinary people chose to ignore both messages and get on with enjoying their time off. Those that did tune in didn't hold their breath.

* * *

Sophy Ridge gets her money shot from the Maybot

8 JANUARY 2017

And then there were three. Not so long ago *The Andrew Marr Show* was the only Sunday politics TV game in town. Then came *Peston on Sunday* and now Sky has joined the party with *Sophy Ridge on Sunday*. Clearly the television bosses have worked out that in the current climate there are more than enough politicians willing to shoot their mouths off to go round. Or not in the case of Theresa May, who has made a career out of saying nothing very meaningful in her first six months in charge of the country.

'We'll be speaking to the prime minister later in the programme,' said Ridge at the start of the show, trying to make this sound as if she was saving the best till last and not that she expected it to be an uphill struggle and a futile encounter for all concerned. Getting the first live interview of the year with the prime minister is supposed to be a major coup. But for the Maybot, everyone makes an exception.

Instead, Ridge chose to warm up with a pre-recorded interview with Labour's Tom Watson in a cafe in his West Bromwich constituency. 'When's the last time you spoke to Jeremy Corbyn?' she asked. 'Ooh, let me think,' Watson smiled. 'We did text one another yesterday about the death of the art critic John Berger.' Watson paused to smile some more. 'And he did ring me last year to wish me happy Christmas.' Good to know that the Labour leadership has their fingers on the big issues of the day.

'So what's Labour's policy on immigration?' Ridge asked. Watson smiled and scratched his head before admitting that he didn't really know. The next time he saw Corbyn – sometime in March, probably – he'd make a point of asking him. 'Labour has a cauldron of ideas,' he said, his smile now veering towards a smirk. Though none that appeared to come readily to mind. He ended by singing his own version of the new Labour anthem: 'Things can only get worser.'

After a quick roundup of the newspapers – a staple of every Sunday politics show – and another pre-recorded clip of Ridge looking thoroughly miserable trying to find anyone in Boston, Lincolnshire – the most pro-Brexit town in the country – to talk to her, it was time for the Maybot. Ridge immediately went on the offensive, observing that immigration had gone up while she was home secretary so why should anyone trust her on Brexit? Which was more important: reducing immigration or remaining part of the single market?

The Maybot whirred into inaction. We were going to take back control of our borders and get the very best deal for Britain because Brexit meant Brexit and the people had spoken. Understandably, Ridge looked a little puzzled by this and tried asking the same question again. This time the Maybot insisted the issue wasn't that binary

– which will have come as news to the other 27 members of the EU for whom it is precisely that – but that we were still going to take back control and get a deal that would be good for Britain as well as the EU.

'You seem to be saying that controlling immigration is more important than staying in the single market,' Ridge said, trying to make some sense of what she had heard. The Maybot looked horrified that any of her answers could have been clearly interpreted and was at pains to reiterate her confusion. This brought the interview seamlessly on to Sir Ivan Rogers' resignation letter that claimed the government was muddled over Brexit. 'Nothing could be further from the truth,' the Maybot rattled. Knowing you're muddled is a sure sign that you're not muddled.

Sensing, like so many before her, that she was getting nowhere, Ridge moved on to the crisis in the NHS. 'There isn't a crisis,' the Maybot snapped. People dying on trolleys in A&E were just scare stories. Clearly this was one issue that was binary. Ridge didn't get much further with the prime minister's new vision for the 'shared society'. How was it different from David Cameron's 'big society' that got quickly dropped? Cameron's society was big and hers was shared. The state would intervene when the state thought it was a good idea and not when it wasn't. OK?

Then Ridge went for the unexpected. 'How do you feel about Donald Trump talking of grabbing women by the pussy?' Ridge asked. The Maybot turned pale, her eyes narrowed and her head looked as if it was about to spin projectile vomit all over the studio. Bugger it. Marr and Peston would never have asked that question. She should have stuck with them, after all. It had been a struggle, but Ridge had got her money shot in the end.

Maybot fails to channel happier times with theme of social injustice

9 JANUARY 2017

Breathe in . . . and relax. When your first six months in office haven't gone at all to plan and everyone is accusing you of having done nothing except get muddled, there's only one thing for it. Go back to the beginning. When Theresa May made her first speech outside Downing Street after becoming prime minister, she spoke passionately about social injustice. About how if you were poor you would die nine years earlier, how if you were black you were treated more harshly by the criminal justice system. That kind of thing.

She hadn't really meant any of it, which was why she hadn't got round to doing anything about it, but it had gone down well enough with the punters. So for her first major speech of the new year, the Maybot chose to channel a happier time. A time when people still believed in her. The previous six months had never happened. In her mind's eye she was still standing outside Downing Street on a July afternoon and not stuck in a room in the Royal Society building off Pall Mall in January talking at the Charity Commission AGM.

'If you're poor,' she began, 'you are likely to die nine years earlier; if you're black . . .' Inevitably, it was all said with much less commitment than before. It's hard enough making the same pitch twice, and nearly impossible when the strain of office has given you the appearance of a computer-operated zombie. Inevitably, too, it didn't go down quite so well the second time round, especially in front of an audience of rightly sceptical public sector and charity professionals.

Still, on the up side, at least she wasn't having to pretend to sound intelligent when talking about Brexit. Anything but that. Cheered by that thought, the Maybot went on to explain her ideas for a 'shared society'. Though 'explained' and 'ideas' might be putting it a little strongly. More like a few thoughts she had come up with over the weekend to take her mind off Brexit. Basically the shared society was a bit like David Cameron's big society, only shared rather than big. And just as likely to be forgotten within a couple of months.

The Maybot wanted to help the little people. Not the really little people who were so broke and so fed up they would never dream of voting Conservative. She wanted to help the not quite so little people, those who were just about managing. Once again, a pall of déjà vu descended on the room.

Perhaps not surprisingly, the Maybot got a little bit confused at this point. After blaming Cameron for most of the mess the country was in – she seemed to have forgotten she was home secretary in his government and had plenty of opportunities to make her voice heard during the referendum campaign – she then went on to give almost the exact same speech about mental illness that her predecessor had given the year before. She even promised to spend the same £1 billion that Dave had promised to spend but had never quite got round to delivering.

Having struggled to the end of the shared cashless society, the prime minister reluctantly took a few questions, most of which were about Brexit. The Maybot clanked with frustration. Hadn't she just spent half an hour trying to get off the subject? 'The pound has decreased in value by 1% overnight since you said you were pursuing a hard Brexit,' observed one reporter. 'Had the City got things wrong or had she?'

'Neither,' she snapped testily. The problem lay with the media, which insisted on reporting her accurately. It was time everyone stopped taking everything she said so literally. After all, it must be obvious to everyone that she really doesn't know what she is doing, so it is grossly unfair to misrepresent her as someone who isn't completely winging it. 'I have been completely clear that Brexit means exactly the same thing I said it meant a couple of months ago,' the Maybot added gnomically. Though not necessarily what she said it meant the day before.

It's not EU, it's us: Maybot outlines Brexit divorce plan

17 JANUARY 2017

For the past six months Theresa May had been trying and failing to conjure some sort of credible Brexit plan and all she'd got for her efforts was a load of flak for being clueless. Eventually, she just got fed up. If she couldn't come up with a deal she believed in, then she would come up with one she didn't. She knew it would be economically disastrous for Britain to leave the single market, but if that's what it took to get everyone who had voted to leave the EU off her back then so be it.

Now to share her moment of enlightenment with the rest of the world. And where better than in one of the gilt-lined state rooms of Lancaster House that had doubled up as Buckingham Place for the Crown? As the room began to fill – cabinet ministers and EU ambassadors in the centre rows, hacks and No. 10 apparatchiks to the side – a lone TV screen displayed the message 'Plan for Britain'. And nothing else. It was a start, I suppose.

The Maybot slid into the room almost unnoticed and took several deep breaths. She glanced down at her notes. There at the top in capitals was 'TRY TO SOUND HUMAN. WE'VE GOT TO NEGOTIATE WITH SOME OF THESE PEOPLE LATER'. Easier said than done, but she'd do her best. She'd start by trying to soften everyone up.

'It's not you,' she said, looking directly at the ambassadors. 'It's us'. Britain had simply outgrown the EU and no longer wanted to be constrained by sleeping with only 27 partners. Britain wanted to go and shag the rest of the world. We had asked for an open marriage and the EU had said no, so a divorce was inevitable. But no one should panic. Britain wasn't leaving Europe. Much as we'd like to if that was geographically possible.

Understandably, being dumped live on global TV didn't go down terribly well with the visiting ambassadors. There was a lot of head-shaking and whispering. The Maybot tried not to catch their eye and moved on to the part of her speech marked 'Global Britain'.

'The referendum was a vote for a new global Britain,' she insisted, hoping that no one would remember that many people who had voted to leave the EU wanted as little as possible to do with the rest of the world. Too full of foreigners stealing our jobs. 'Global Britain, global Britain,' she repeated. If she said it often enough, someone might believe it. Even if she didn't.

Having got the pleasantries out of the way, the Maybot moved on to her 12-point plan. She wasn't entirely sure why there were so many points in the plan, as most of them were just vague promises to clarify things that hadn't yet happened, but her advisers had told her that cutting the plan down to just three points – 1. Get out the single market. 2. Get out the customs union. 3. To hell with the lot of you – wouldn't play out that well.

'Global Britain, global Britain,' the Maybot droned. If in doubt, return to the leitmotif. Britain wanted to be really, really global – Michael Gove had promised her that Donald Trump was dead keen on coming to a win-lose deal with the UK – but she was keen to extend the hand of friendship to the little old EU. So if the EU was prepared to give us everything we wanted in the negotiations, then we might just continue to trade and talk to it.

But it was up to the EU to show willing. 'It would not be the act of a friend to treat us punitively in the negotiations,' she said. Several ambassadors had to pinch themselves at this point; it hadn't been the act of a friend to vote to leave. But the Maybot had one more sting in the tail. 'No deal for Britain is better than a bad deal for Britain,' she added. Her way or no way. If the worst came to the worst, Britain could always settle for WTO rules and become a tax haven.

'The country is coming together,' she concluded. One last white lie wouldn't hurt. She knew it had never been so divided, and the divisions were only likely to get deeper now she had declared her hand for a hard Brexit that she hadn't voted for and didn't want. But people had demanded clarity and she'd given it to them. There was almost no applause at the conclusion of her speech but for once she didn't care.

Somehow she felt a little lighter. A weight was off her shoulders.

* * *

Theresa May's Lancaster House speech prompted an unexpected display of sweetness and light on the Conservative benches at prime minister's questions the following day. It was as if the divisions of the past six months had never happened, as the Eurosceptics joined ranks with the Europhiles to praise the prime minister's brilliance.

Anna Soubry, normally Theresa May's bête noire, could only marvel at the clarity and leadership the prime minister had provided and just hoped that she could write down her 12 objectives so parliament could venerate them as holy relics. Alistair Burt, another erstwhile troublemaker, was overwhelmed by how constructive the speech had been and prayed that God would grant her even more constructiveness in the months and years ahead.

Even the doggedly pro-EU Ken Clarke had only the most mild of rebukes for Theresa, saying it would have been nice if she had come and told parliament about what she was going to say first. The days when the Tories washed their dirty linen in public appeared to be at least temporarily over.

Sensing her moment, the Maybot chose to ignore any inconsistencies in her position and chose to repeat selected highlights of her Brexit speech. Only this time adding a little more hyperbole. 'We are a global Britain,' she had said. 'We are a more outward looking, more tolerant country. We embrace everyone.'

Except the 27 countries of the EU who had been nothing but a burden to us, holding us back in our attempts to make the world a better place for everyone – even those we didn't particularly like and wouldn't want anywhere near our green and pleasant land.

Except also the 12 Supreme Court judges. Well, eight of them. After hearing the government's Article 50 appeal in December, the Supreme Court had delayed delivering its judgment till the end of January. Not so much because there had been a great deal to think about, but more because it didn't want anyone in government to think it hadn't spent enough time deliberating over its verdict. And besides, what judge could resist a bit of fabricated suspense?

But on January 24th, the Supreme Court duly upheld the original decision of the High Court. The pro-Brexit newspapers were again outraged. The government wasn't best pleased either.

* * *

David Davis sees Article 50 defeat as a win in his alternative facts narrative

24 JANUARY 2017

As Lord Neuberger delivered the Supreme Court judgment, the attorney general, Jeremy Wright, slowly shook his head. It must have been the Pavlovian response of a lawyer used to losing most of his cases, as Wright was the only person in court who appeared to be surprised by the verdict. There again, he had probably been the only person in court to have just received a text from the prime minister that said: 'I gave you one job. ONE JOB.'

A few hours later, David Davis came to the Commons to explain why the government had never believed the 'Enemies of the People' really were the 'Enemies of the People' and had only challenged what the 'Enemies of the People' had originally said to enable the rest of the country to understand why they weren't the 'Enemies of the People'. The alternative facts narrative is catching on in the UK. Michael Fallon on Monday, Davis on Tuesday and no doubt Theresa May on Wednesday.

Everything had gone exactly to plan, Davis insisted, his fists clenched tight, and the government had only tried to avoid letting parliament have a say in the triggering of Article 50 in order to let the judiciary assert its independence. But now the judges had had their say, he was going to take back control by scribbling a few sentences on the back of an envelope to put before parliament. In the interests of national unity, the will of the 52% could not be denied.

'This has been a good day for democracy,' Keir Starmer, shadow Brexit minister, replied, 'and the prime minister was wrong to sideline parliament.' May, sitting next to Davis, had the grace to squirm uncomfortably. Starmer expressed his surprise that the government thought it could pass off a speech about leaving the single market – made to a few ambassadors at Lancaster House – as proper parliamentary scrutiny and then say the whole Supreme Court appeal was a massive waste of time and money.

Davis adopted his best hurt face. The prime minister had never been trying to sideline democracy. Rather she had been trying to mainline it. Her only crime had been to carry out the will of the people by making sure that parliament was not given a chance to take back control of the process of taking back control of parliamentary democracy. He then paused, waiting for a response. None came. If he could get away with nonsense like that he could probably get away with anything.

Buoyed by this thought, he reiterated his position. Losing the appeal had been part of a cunning masterplan and the government could not be more pleased with the verdict. He had already said three times that the 'Enemies of the People' were not really the 'Enemies of the People' and he would continue to do so. The fact that the 'Enemies of the People' had upheld the verdict of the 'Junior

Enemies of the People' only proved how right the government had been to appeal against the original decision because it provided clarity.

What Davis wouldn't do, though, was give any kind of statement that might undermine the government's negotiating position, because that wouldn't be in the national interest. As for what the national interest was, it would all become clear at the end of the Brexit negotiations, as whatever the government managed to negotiate would turn out to be in the national interest. He really couldn't be clearer than that. 'I've already come to the house five times to make statements,' he moaned, 'so you can't accuse me of saying nothing.' No one had the heart to tell him that the reason he had needed to make a fifth statement was both because he'd said next to nothing in the previous four and because the prime minister had picked an unnecessary fight with parliament and the judiciary.

'The split judgment shows the prime minister was right to appeal to the Supreme Court,' said a deranged Iain Duncan Smith, for whom an 8–3 defeat is a moral victory. It could have been worse. They could have lost 11–0. Goal difference is going to count for a lot in Brexit apparently.

Like everyone else, Davis wisely chose to ignore IDS but was then forced to listen to countless MPs from both sides of the house asking him to put a white paper before parliament. 'No,' said Davis. He wasn't going to provide that sort of detail as it would only give people a chance to table amendments. He hadn't taken back control only to give it away again. Rather he was just going to do the bare minimum the 'Enemies of the People' had required of the government. Not that they were 'Enemies of the People' of course.

* * *

It had been a big enough shock for the government when Donald Trump had been elected president of the USA the previous November; if either Theresa May or Boris Johnson had had any indication that The Donald would become one of the two most powerful men in the world they would never have been so rude about him during the presidential campaign. But worse was to follow when it was Nigel Farage who got the invitation to the inauguration and Michael Gove – in the company of Rupert Murdoch – who received the second audience with the president to write a sycophantic profile. The snub to Theresa May could not have been made clearer. Even though she was prime minister, in the president's mind she was only the third most important British politician. Her advisers quickly phoned the White House to secure a meeting.

* * *

Theresa May and Trump: PM shows lengths she'll go to for Britain

27 JANUARY 2017

The body language could hardly have been more awkward as Theresa May and Donald Trump posed for their blind date in front of the bust of Winston Churchill in the Oval Office. The prime minister kept her distance and looked faintly embarrassed, as if it was only just dawning on her that the main reason she was the first foreign leader to meet the US president was because all the others had thought better of it. That and the fact she was a bit desperate. Britain doesn't have as many friends as it used to.

Trump merely looked a bit blank. Perhaps this was because the British prime minister wasn't the woman he had been expecting. All morning the White House had been tweeting that he was about to meet Teresa May, the spelling mistake turning the prime minister into a porn star. The special relationship has always been rather more special to us than the Americans. As the two leaders finally shook hands, the bust of Churchill covered its eyes and begged to be sent back to Britain. Their hands remained uneasily entwined as they walked down the colonnade towards the Palm Room. When Trump started to creepily stroke her hand, Theresa almost retched. She quickly pulled herself together and reminded herself to just think of England. Sometimes you had to take one for the team.

Things had marginally improved by the time Trump and May began their press conference an hour or so later. They could at least look each other in the eye. The president kicked off by saying how honoured Theresa was to be the first leader to the White House and that their talks had highlighted Brexit was indeed a blessing and the special relationship was going to be even more special from now on. More special for whom was left unanswered. His fingers appeared to be moving along his script as he talked. If Trump wasn't exactly presidential, at least he held his narcissism vaguely in check.

Theresa appeared to have got rather more out of the talks than Trump. After promising Trump a ride in a horse-drawn coach with the Queen and a game of golf in Scotland, she had definitely heard the president say he was '100% behind NATO' even though he thought it was obsolete. She also said she would bully other EU leaders into paying more cash into NATO. Like they were going to listen to her. The prime minister also seemed to think the prospects were good for the US selling us chlorinated chicken and taking over the NHS and of course she understood that the US might want to terminate with 30 days' notice.

Things became rather more interesting when the BBC's Laura Kuenssberg asked the awkward questions about torture, banning Muslims and being untrustworthy that

Theresa had somehow forgotten to ask. 'There goes that relationship,' said Trump. He was only half joking. The president went on to say that he was all for waterboarding but if his defence secretary decided not to use it that was OK by him as he could always be waterboarded into changing his mind.

'There will be times when we disagree about things we disagree about,' said Theresa trying not to sound too fawning while not wanting to upset her host. Her need was far greater than his. The Donald quickly veered off track. 'I've had many times when I thought I was going to like someone only to find I don't,' he replied. 'So Theresa, we never really know about these things.' Theresa didn't look altogether relieved. Not even when Trump immediately contradicted himself by saying he was a people person and could instantly tell if he was going to get on with them.

Theresa closed her eyes and hoped Trump wouldn't get too sidetracked. Her prayers were answered as The Donald returned to Brexit. Brexit was great because he had predicted it and he'd never liked those EU consortium guys because they had once cost him money by blocking one of his deals. 'That's why you needed Brexit,' he concluded.

The last word went to Theresa, who tried to wrap things up in a more statesmanlike fashion by talking about 'ordinary working people', but by now Trump had already got bored and was tucking his speech inside his breast pocket as she spoke. Twenty minutes is about all his attention span can take. Theresa breathed a sigh of relief. It could have been worse. He could have started a war. He could have grabbed her pussy. At least she'd got out with some of her self-respect intact. Not much. But some.

* * *

The Supreme Court ruling meant that the government had to hurriedly cobble together a bill to put before parliament if it was still to meet its timetable of triggering Article 50 before March 31st. In hindsight, allocating two days to a debate on a 137-word bill was at least a day and a half too long. Not because the bill to trigger Article 50 was of no importance, but because it gave MPs licence to rehash old speeches and fight battles that had already been long since won and lost. Due process may be a democratic necessity but it's not always pretty. Or edifying.

'We've already reached the point of no return,' David Davis had growled. The Brexit secretary's voice had been even croakier than usual and he clearly regarded this latest outing at the dispatch box as beyond the call of duty. He still hadn't quite accepted that he could have saved himself a lot of time and effort had the government agreed to publish a white paper six months ago. But as he was there, he would briefly recap. Brexit meant Brexit and that was that.

The debate that followed the Brexit secretary's opening speech was something of a damp squib, as Jeremy

Corbyn had already made backing the government a three-line whip for Labour MPs. Verbal wallpaper to fill the time before the vote. Heaviest of hearts / gravest of misgivings / Britain never, never will be slaves – delete where applicable. Only those Labour Remainers, such as Stella Creasy, Chris Bryant and Mary Creagh, who were planning to ignore their party line, managed to inject any real passion. It's so much easier to sound sincere when you actually believe in what you're saying.

Jenny Chapman and David Jones had been no less prosaic in summing up for both sides. Perhaps they had felt they had to live down too much of what had gone before. Jones even managed to lose his place and announce that Britain would be leaving the UK. Now that would have been news.

It was hard to believe this had all been a preamble to what was a historic moment. Only a year ago, Euroscepticism had been a strictly top-shelf activity, limited to a few hardcore fetishists. Now parliament was about to vote in favour of something most MPs knew to be a bad idea. With only the Lib Dems, the SNP, Ken Clarke and a few Labour refuseniks voting against, the bill was passed by an overwhelming majority.

The following week, Labour tabled an amendment to the Article 50 bill calling for a meaningful vote to any final deal and during the committee stage reading in parliament, Brexit minister David Jones made an announcement. After careful consideration of the possibility that there might be rather too many Tory rebels for comfort, the government was prepared to allow parliament a vote on Britain's withdrawal from and future relationship with the EU before the deal was voted on by the European parliament.

Keir Starmer welcomed this as an important concession, but Jones was quick to put him right. He was calling it a concession because he needed to be seen to be making a concession. But to avoid any confusion, this concession was most definitely not a concession. What it was was a meaningless diversion to allow some of the Tory Remainers to feel a bit better about themselves when they voted with the government later in the afternoon.

'So this isn't a concession, is it?' Ken Clarke had observed.

Jones had smiled wanly. He didn't know how to put this any plainer. All that was on offer was a vote on a deal or no deal. With just a couple of minutes to debate it. Just as Theresa May had promised in her Lancaster House speech. Parliament could either accept whatever bad deal the government managed to negotiate with the EU or it could jump off a cliff by going straight to World Trade Organisation rules.

'But for a vote to be meaningful,' Labour's Chuka Umunna pointed out, 'parliament must be able to send the government back to the EU to renegotiate'.

A howl escaped Jones's lips. That was precisely why there wasn't going to be a meaningful vote. How many more times did he have to repeat himself? If he had made a concession, and Jones was certain he hadn't, then it had been to allow a meaningless meaningful vote. A vote whose only meaning was in its absence of meaning. And yet it was a vote that was enough to assuage the consciences of any Tory Remainers who had been considering voting against the government.

With every other arm of government seemingly either confused or making up policy on the hoof on a daily basis, it was perhaps no surprise that chancellor Philip Hammond appeared to be doing the same in his spring budget. That, at least, was the kindest explanation of why he might have forgotten the Tory party manifesto promise not to raise national insurance contributions.

* * *

The Undertaker's budget brings death, taxes then a crazy kamikaze attack

8 MARCH 2017

He'd come to praise the economy. But while he was there he might as well also bury it. They didn't call him Phil 'The Undertaker' Hammond for nothing. This was to be the Undertaker's last spring budget. Just as well, as he didn't really have anything much to say. Not that it would stop him from taking his time in not saying it. Seldom has a chancellor been on his feet for so long and said so little.

With prime minister's questions over, the Undertaker lurched towards the dispatch box. 'The economy has shown robust growth, the deficit is down and the labour market is strong,' he said. At this point, he paused and scratched his head. If everything was going so well, why the hell was the country leaving the EU? Best not to ask that sort of question. Way too far above his pay grade. If the country wanted to bury itself, the least he could do was give it a proper send-off.

'We can't rest on our past achievements,' the Undertaker went on. No one could argue with that, as there hadn't been many to rest on. The growth forecasts might have been better than last autumn but they were still well down on this time the previous year. Ten minutes in and with nothing of any interest said, a few heads started to go down on the Tory benches.

Time for a gag. The bereaved always appreciated a little humour. Nothing too funny, mind. No danger of that. 'The last Labour government,' he said. 'They don't call it the last Labour government for nothing,' he explained. Theresa May squeezed out a forced laugh. The Undertaker beamed with pleasure. He hadn't had that good a reaction since his double cremation gag in Croydon back in 1998.

The Undertaker droned on. And on. He liked the sound

of his own voice even if no one else did. A funeral address had never been so funereal. Never before had the rich paid so much in tax. The Tory backbenchers didn't look entirely pleased by that, but the Undertaker was quick to reassure them. Think about it this way: the rich had been earning more and more while everyone else was going broke. A celebration of inequality.

Social care. He supposed he had better say something even though there wasn't much to say. Who cared if a few old people croaked? More work for funeral directors. Every cloud and all that . . . Sure he'd bung them an extra £2 billion. But only over three years. It wasn't nearly enough but the opposition was too feeble to do anything about it. Besides, it would serve the country right. A hard Brexit was a gonna fall and the body count would rise.

But the Undertaker was damned if he was going to pay for other people's funeral costs. The self-employed could pay for the social care bill. It's about time those freeloaders paid their fair share of national insurance contributions. Theresa urgently tapped him on the shoulder. 'Psst,' she said. 'We promised not to increase NICs at the last election.' The Undertaker picked up a copy of the manifesto and crossed out the relevant section. Sorted.

And that was it. Everything was pretty much as expected. Wage increases were offset by cuts to in-work benefits and rising inflation. The continuous beep of the cardiograph was music to his ears. The country was flatlining. As was the Commons after 50 minutes of nothingness. One last gag and he was done. 'One day we might have driverless cars,' he observed. 'And the party opposite knows all about being driverless.' Some MPs nearly laughed. It was a toss-up whether they would rather be in a car with no driver or a hearse being driven by a dozy funeral director.

Responding to a budget is one of the toughest gigs of the parliamentary year and most opposition leaders hastily scribble a few notes while the chancellor is on his feet. Jeremy Corbyn made none. He had his speech already written and he was determined to stick to the script. Only it was more of a rant than a speech.

'This was a budget of utter complacency,' he shouted, his voicing ramping up to volume 11, 'utter complacency about the crisis facing our public services . . .' The louder he shouted about the evils of the Tories without bothering to address a single issue raised by the chancellor, the less anyone was inclined to listen.

Corbyn didn't even appear to have clocked that the chancellor had broken an election pledge. Rather, it was a deranged kamikaze attack in which the Labour leader was the only fatality. Still, it was an achievement of sorts. It's not every day that someone manages to empty the chamber faster than the Undertaker.

Philip Hammond digs deep as he explains his NICs U-turn

15 MARCH 2017

The phone call had come through just after eight in the morning while Phil 'The Undertaker' Hammond was eating breakfast. It was the prime minister ordering him to bury Class 4 NICs. He had tried telling her that doing a U-turn on your only real budget measure less than a week after it had been announced made him and the government look hopelessly incompetent, but Theresa wasn't having any of it. The Tory backbenchers were on her back. The *Daily Mail* was on her back. And now she was on his back.

Six hours later the Undertaker rather sheepishly arrived in the Commons to try to explain how it was that, though he still absolutely stood by his budget because it was his budget that was his, he now wanted to fundamentally change it because although he hadn't broken any promises in the Conservative party manifesto, as that's not the sort of thing he would ever dream of doing, he had in fact broken the promises he had made in the Conservative party manifesto.

It had been absolutely right to raise NICs and that's why he wasn't doing it. And no, before anyone asked, he hadn't worked out how to fill the £2 billion black hole that had just opened up in the country's finances. Give him another six months. Maybe changing to one budget a year wasn't such a good plan after all.

Not even Jeremy Corbyn's hapless efforts to take advantage of the government's uselessness at prime minister's questions could cheer the Undertaker up. He was a proud man. A vain man. A man who still kept his Mr Funeral Director of the Year 1978 award on his mantelpiece. A man who sensed he was now living on borrowed time. A man both dead and undead. He'd screwed up at the first hurdle. Another cock-up and he would be a goner.

The Undertaker only knew one way. When in doubt, keep digging. 'It wasn't either myself or the prime minister who realised we had broken the Tory party manifesto which he hadn't broken,' he said. 'It was the BBC's Laura Kuenssberg.' Great. So neither the chancellor nor the prime minister had a clue what was in their own manifesto? 'No, no, no,' said the Undertaker hurriedly. 'What I really meant was that when I had leaked that we were planning to break our promises to the media ahead of the budget, no one had raised an eyebrow, so I thought it was all going to be OK.'

By now only the Undertaker's head was still showing above the grave he had just dug for himself, and Labour backbenchers were queuing up to deliver the coup de grace that their front bench had failed to land. The Undertaker's misery was only compounded by the support from his own benches. Being told he was a brave man could only mean one thing. There were a lot of MPs who were mighty pissed off they had spent the past week defending a crap policy that had now been ditched. The Undertaker made one last appeal. 'I want to restore the faith and trust of the British people,' he mumbled, dimly aware he had just done the exact opposite, 'especially as we embark on the process of leaving the EU'.

He might not have been quite so quick to say that if he had listened to David Davis give evidence to the Brexit select committee earlier in the day. Therapists often like to see their patients first thing in the morning because they are then at their most undefended and aren't awake enough to portray themselves in a good light. It certainly worked a treat with the Brexit secretary, who appears to have aged five years in the last few weeks. Though that could have something to do with the large quantities of sodium pentothal he had just taken.

The treasury select committee chair, Hilary Benn, warmed up with a bit of free association. Did no deal mean WTO tariff barriers? 'Yes,' said Davis. Would there be border checks between Northern Ireland and the rest of Ireland? 'Yes.' Would the EU/US open skies agreement be dead in the water? 'Yes.' Would we lose passporting rights of financial services? 'Yes.' Did this mean that the foreign secretary was idiotic to say that dropping out of the EU on WTO terms would be fine? 'Yes.' Whoops. He had just landed Boris in it. Still, Boris wouldn't have thought twice about knifing him.

Benn then went for the throat. Had Davis made any

calculation of the exact costs of leaving the EU on WTO terms? 'God no,' said Davis breezily. 'I know how it's going to work out. I just haven't quantified it.' Every member of the committee – even the Leavers – stared into the abyss. Davis had just admitted the government was saying no deal would be better than a bad deal when it didn't even know the cost of no deal. A parish council wouldn't get away with that level of unaccountability. Davis shrugged. There was something liberating about telling the truth. Why not let the country know that the chancellor hadn't a clue about the economy and Brexit was heading for the rocks? It wasn't as if there was an effective opposition to stop them.

* * *

After attending an EU summit in the middle of March, Theresa May came back to parliament to give a statement on how the weekend had gone. But as almost nothing had happened and she was sent home while the other 27 EU countries discussed Brexit, she didn't have a lot to say. She rifled through her notes, trying to fill in time. 'I did call on the EU to complete the single market in digital services as that would be in the UK's best interests,' she said. Only someone with a synaptic disconnect could have remained oblivious to the irony of urging everyone else to sign up to something she was committed to leaving. But Theresa effortlessly sank to the occasion.

Jeremy Corbyn asked her how the Brexit divorce proceedings were going. 'Don't call it a divorce,' Theresa May had replied crossly. 'Brexit isn't a divorce.' She was right. A divorce implies two parties more or less amicably agreeing their terms of separation within a couple of years. Brexit was going to be messier than that. Much messier. The chances of reaching a mutually satisfactory financial settlement and trade agreement in that short a time were almost non-existent.

This didn't entirely impress the SNP. Scotland had voted by a large majority to remain in the EU and the SNP weren't keen on being dragged out of the EU as part of a job lot with the rest of the UK. And with parliament only being allowed a meaningless, meaningful vote on a final Brexit deal, Nicola Sturgeon, the SNP leader, had called for Scotland to be allowed a second independence referendum.

This was something Theresa was unwilling to grant. The Scots had had their chance and blown it. Now they just had to suck it up until she was ready to give them one. Or not. She hadn't decided yet. Either way, Scotland was coming out of the EU. End of. What it had to accept was that it was better off being part of a union. Even if the United Kingdom was better off out of one. Logic had never been the prime minister's strongest suit.

* * *

Maybot stuck on repeat as Sturgeon lets rip over referendum

16 MARCH 2017

'Now is not the time for a second independence referendum,' said Theresa May, tilting her head to one side like a patronising Princess Diana and fluttering her eye-lids over the shoulders of ITV's Robert Peston into what she imagined was the hearts of the Scottish people.

At times like this Theresa felt it was her destiny to be the Queen of Unionist Hearts. Even though she still wasn't sure where Scotland was exactly. But she somehow just knew she would love it if she ever got round to finding out. The Scots stared back impassively.

'Then when is the right time?' enquired Peston reasonably.

'Now is not the time.'

Peston tried again. 'Can we be clear about when you do think is the right time?'

'Now is not the time.' A virus had re-infected the Maybot and she was stuck on repeat.

'Yes, I get that, but . . .'

'Now is not the time,' said the prime minister, unaware she was turning her bad week into a worse one.

'So what you're saying is . . .'

'Now is not the time.'

Over in Holyrood, the Scottish first minister, Nicola

Sturgeon, was only too happy to agree. Now was not the time to hold a second referendum. But sometime late next year when the Scots had had a chance to see how badly they were going to be screwed over Brexit would be.

The promised UK consensus that the prime minister had offered on any Brexit deal had already been relegated to a few text messages: 'Soz. We R leaving the single market' and Sturgeon didn't trust Westminster not to sell her country even further down the river.

First minister's questions in Scotland is an altogether more enlightening affair than prime minister's questions down south. Not least because serious questions get asked. And answered. It helps that the two main adversaries, Sturgeon and Conservative Ruth Davidson, are rather sharper than their UK counterparts – not difficult for Davidson as Jeremy Corbyn hit a new low at PMQs the day before by even forgetting to ask a couple of questions. It's also a major plus that the rest of the chamber manages to listen without sounding like a Bash Street Kids school reunion. When each speaker has finished talking, there is a round of applause. Or silence. It's disconcertingly polite.

Davidson opened by asking whether Sturgeon thought it was the right time to call for a referendum when Scottish schools were in such a mess. The first minister eyed her up. A civil question deserved a civil answer. Yes, there were problems in schools and she was doing her best to deal with them but that didn't stop her multi-tasking in the national interest.

'Is it not true, though,' said Davidson, 'that independent forecasts suggest independence would put Scotland £11 billion in the red?'

This was Sturgeon's moment to let rip. Ever so nicely, of course. The reason Scotland was running a deficit was because it had been under the control of the Westminster purse strings for so long. Surely it was time for Scotland to see if it could do better on its own, rather than risk being made even worse off by a hard Tory Brexit? And if they couldn't then at least there would be the consolation of knowing the pain was self-inflicted?

Davidson kept going. She rather had to, as she's the last politician standing in the UK between Scotland remaining in the UK and declaring a Unilateral Declaration of Independence. Unlike in the last independence referendum, Labour is now dead in the water in Scotland and the appearance of May on the campaign trail would send voters running into the arms of the SNP.

'I choose to put this parliament first,' said Davidson.

Bad move. Sturgeon quickly reminded her opposite number that she had a far higher share of the vote than Theresa – even taking into account the dodgy counts in Thanet and elsewhere – and had been elected on a manifesto that had promised a second referendum. 'So I issue a direct challenge,' she concluded. 'If next Wednesday, the Scottish parliament votes for a second referendum, will the Tories respect the will of this parliament?' Sod it. A party that lived by 'The Will of the People' could also die by it.

Back in London, Theresa experienced a glimmer of hope. She may have just made a second independence referendum inevitable. But at least she'd given herself an even chance of delaying it until the Scots were completely penniless.

* * *

On the afternoon of 22 March a 52-year-old British terrorist, Khalid Masood, drove into a group of pedestrians on Westminster Bridge, killing four people and injuring 49 others. Masood then ran through the main gates of the Palace of Westminster, where he killed PC Keith Palmer with a machete before being shot dead by armed police. The following day, parliament convened to hear the prime minister make a statement on the terrorist attack. In the past the prime minister had often managed to misjudge the big occasions, appearing to be lacking in both grace and warmth, but this time she was note-perfect: calm and informative; dignified with a hint of steel.

At the very same time the terrorist attack was taking place in London, the Scottish parliament in Holyrood was debating whether it should call for a second independence referendum. Once news of the scale of the attack reached Holyrood, the debate was halted. Just under a week later, it was reconvened and the tone of the debate was noticeably more conciliatory, with Nicola Sturgeon emphasising shared values, democracy and differences of opinion that were sincerely held.

Sturgeon also tried to keep things short and sweet. She had tried – my God she had tried – not to call for another independence referendum by begging Theresa to come to some kind of deal with Scotland. But Theresa had repeatedly snubbed her, refusing to even talk about the implications of Brexit for Scotland.

The Scottish Conservatives accused the SNP of political opportunism but the SNP had the necessary majority to win the vote by 69 to 59.

The following day, two days earlier than the prime minister's self-imposed deadline of March 31st, Tim Barrow, Britain's ambassador to the EU, delivered a letter formally triggering Article 50 to Donald Tusk, president of the EU Council.

* * *

End of the affair: May finds breaking up with EU is hard to do

29 MARCH 2017

In his Brussels office, President Tusk ripped open the envelope. 'Dear Donald, hope you are well . . . blah, blah . . . the people of Britain have voted . . . blah, blah . . . it's not you, it's me . . . blah, blah . . . I really want to remain friends, but right now I need some space . . . blah, blah . . . I know I can't expect to have my cake and eat it but if there was any chance of me having my cake and eating it, I wouldn't say no . . . blah, blah . . . joint custody of the kids . . . blah, blah . . . Love from Theresa.'

Tusk scrunched the letter into a ball and tossed it into the bin. It was almost exactly as he had expected. Polite, verging on the over-familiar, in places; rude and a bit threatening in others. Nothing he couldn't deal with quite comfortably. He would give as good as he got over the coming years. But what did surprise him was how emotional he felt. 'We already miss you,' he muttered. 'Thank you and goodbye.'

Back in London, Theresa May was also having a wobble. She had expected to feel nothing but relief at triggering Article 50. No more pretending to listen to the whines of the hardcore Eurosceptic fanatics on her own benches. No more pretending to take any notice of the Scots. No more pretending to pretend that she knew what the hell she was doing. Instead she felt nothing but a sense of sadness and anti-climax. Sadness that she was finally leaving something she would actually quite miss. Anti-climax because after all the build-up she was left with a gnawing sense of emptiness.

The prime minister tried to keep her emotions in check as she delivered her statement on the triggering of Article 50. 'It is a plan for a new deep and special partnership between Britain and the European Union,' she said. 'A partnership of values. A partnership of interests. A partnership based on cooperation in areas such as security and economic affairs.' BFFs. Kiss, kiss, kiss.

A few Tory backbenchers began to look a little uncomfortable. They had come decked out with Union Jack ties, iPad covers and hair bows in anticipation of a glorious celebration of the nation's liberation from the jackboot of Europe. What they were getting was more of a love letter. The longer Theresa was on her feet the more it sounded as if she thought Brussels was heaven on Earth and was begging to re-join the EU as soon as possible.

From time to time, Theresa's rose-tinted fantasies got the better of her. Claiming that Britain would be stronger, fairer and more united, while Northern Ireland is in deadlock, the Scottish parliament has just voted for a second referendum and the Welsh are becoming steadily more disillusioned, indicated a loosening grip on reality.

Nor did she do herself any favours by saying: 'More than ever, the world needs the liberal democratic values of Britain.' The Lib Dems couldn't believe their luck at getting a shout out from the prime minister. Theresa didn't even notice the irony of what she was saying and merely repeated herself after a long interruption for laughter. She was lost in a personal grief.

Jeremy Corbyn surprised no one by responding to the statement that he had expected the prime minister to make rather than the one she had made. He went off on a lonely riff about hard Brexit and bargain basement tax havens, seemingly unaware the prime minister had already committed herself to meeting Labour's six Brexit tests. She wouldn't, of course, but that was beside the point. What mattered was the EU love-in. On days like this the Labour leader redefines the meaning of the word mediocre.

Most of the Tory Eurosceptics – with the exception of Bill Cash, for whom the only good German is a dead German – more or less managed to keep their triumphalism in check. There would be plenty of time for crowing in the days ahead. All that mattered was that the letter had been sent; whether it was a disaster or not was neither here nor there.

Theresa's only difficult moments at the dispatch box came when Angus Robertson pointed out that Scotland had voted to remain and that Theresa had broken her promise to agree a deal with the Scots. For the only time, Theresa looked flustered.

'My constituency voted Remain,' she said. Comparing a country to a constituency was not the brightest move. Realising her mistake, she went full Maybot and repeated, 'Now is not the time for a second referendum' over and over again. The SNP sniffed blood. Now might not be, but sometime soon might well be. The clock was ticking. 730 days and counting.

David Davis: the UK's secretary of state for badly needing a lie-down

30 MARCH 2017

Hanging on the wall of a disused office in the Department for Exiting the European Union there must be a picture of a David Davis who is getting visibly younger by the day. The real Brexit secretary is on the opposite trajectory. Six months ago he was fit, active and young for his age; now he looks done in. Not done in as in a bit tired. Done in as in completely knackered. On his knees. His eyes have developed huge bags beneath them that not even a skilled makeup artist could conceal, his face has sunk and reddened and he moves more slowly. If not always more deliberately. At this rate he will be a shell by the end of the year. A hollow head for a hollow crown.

Sometimes you have to be careful what you wish for. Brexit was meant to be the crowning achievement of Davis's career. A personal and national liberation from all those Brussels bureaucrats who had made his life a living hell for the past 44 years. Not that he could quite remember what it was they had done that was so bad, but he was sure they must have done something simply because they weren't British.

But then Theresa May messed things up completely by putting him in charge of all the complicated Brexit stuff that he had never really properly understood. To make things worse, with both Liam Fox and Boris Johnson reckoned to be total liabilities by the prime minister, he was always the one sent out to do all the tricky media stuff. So by the time he had spent several hours touring the TV studios and radio stations trying to explain why the threat to let foreigners die if they didn't do a deal wasn't really a threat, he was half-asleep by the time he had to present the great repeal bill to the Commons.

'Exciting opportunity . . . going forward . . . embracing change,' he said in a barely audible monotone. It was one thing to remember Theresa's advice of accentuating the positives but beyond him to do so as if he meant it. 'The great repeal bill will,' he continued, searching frantically for the right words. 'Will . . . will . . . make all EU law UK law.' What happened after that was anyone's guess. He didn't have a clue how much EU law was already UK law, what laws the government might want to change, how they would be changed or who would get to do it. But no one would be too interested in those sort of details, surely?

Davis looked horrified when Keir Starmer homed in on precisely those areas. Labour's shadow Brexit secretary can sometimes be rather awkward at the dispatch box but as a lawyer, today he was in his element. Starmer wanted to make sure no fundamental rights were going to be revoked and that the government wasn't looking to do away with some laws on the sly without proper parliamentary scrutiny.

'I ask the minister to look again at this,' said Starmer. Davis almost burst into tears. Please no. Anything but that. He'd given it all a quick glance before he had come out and surely that was enough.

'We'll put things right if we've missed anything,' said Davis. 'I promise.' Scout's honour. Though he couldn't extend that promise to necessarily allowing MPs to have a vote on anything because some of them – he was looking at the SNP now – might try to block it. After all, what was the point of the government making life any more difficult for itself? Things were going to be tricky enough now that Article 50 had been triggered and the Germans and French had the upper hand without having to worry about attacks on the home front.

Nick Clegg began by congratulating the Brexit secretary for not pandering to the hard-line Eurosceptics in his party. Davis looked concerned. Praise from a committed Europhile was not the response he had been expecting. 'I'm delighted to see you have kept the jewels in the EU crown, such as the working time directive. But why, if we are to keep them, are we going to the effort of leaving the EU anyway?' Davis said nothing. All that was well above his pay grade.

A few Eurosceptics woke up at this point and started mumbling about 'the ghastly EU' and 'a glorious return to parliamentary sovereignty'. Albeit a return without a return to parliamentary democracy. Davis happily acknowledged their wisdom. We were getting rid of all

European Court of Justice legislation, he insisted. Apart from the bits we weren't, which would be called the British European Court of Justice legislation. Satisfied everyone was now as confused as him, Davis went for a lie-down.

* * *

During the nine months that she had been in office, Theresa May had repeatedly insisted that she wouldn't be calling an early general election. 'There should be no general election until 2020,' she had said in June 2016 as she tried to present herself as the continuity candidate in the Conservative leadership race.

In September 2016, she had told Andrew Marr, 'I'm not going to be calling a snap election. I've been very clear that I think we need that period of time, that stability to be able to deal with the issues that the country is facing and have that election in 2020.' The following month, she gave an interview to the Sunday Times *in which she said a general election before 2020 would bring unwanted instability to the country.*

Amid mounting speculation in March 2017 that the Tories would be fools not to take advantage of their 20-point lead over Labour in the opinion polls to win themselves a landslide majority, No. 10 categorically denied that Theresa May was planning on calling an early election on at least two occasions. 'It is not going to happen,' the prime minister's spokesman said.

But then, on the first day back after the Easter break, it did happen. The official version was that Theresa May had been reluctantly talked into it by her husband, Philip, while they had been walking in Snowdonia over Easter. The rather more believable version was that she had had her arm twisted by her two advisers, Nick Timothy and Fiona Hill, along with some of the more hawkish members of her cabinet, and had come round to their way of thinking that an opportunity to kill off the Labour party for a decade or longer was just too good to pass up.

* * *

Dead-eyed Theresa May puts the Tories' interests first

18 APRIL 2017

Right to the end, Theresa May was unable to keep to her own timetable. For the past six months, the prime minister had repeatedly insisted she wouldn't be calling an early general election because it wasn't in the best interests of the country. Sometime over Easter, Theresa was blessed with a divine revelation – there are advantages to being a vicar's child – and came to the conclusion that her own party's interests were rather more important than the country's. So shortly before 10 a.m. her office announced that she would be making a statement in Downing Street at 11.15.

Worried she hadn't caught enough people on the hop, Theresa darted out the front door of No. 10 nine minutes early and made a dash for the wooden lectern that had been hurriedly placed outside. She paused to clock her surroundings. Satisfied that comparatively few journalists had made it in time, she got straight to the point. After overdosing on elections in recent years, the country was now going through cold turkey. People were literally crawling up walls out of desperation to vote, and to satisfy their cravings she was going to give everyone another fix on 8 June.

Not that she wanted to be seen as a prime minister who didn't keep her word. The problem was the opposition. They were doing the wrong thing by opposing her. Never mind that they weren't being very effective, the problem was that they existed at all. They were a nuisance. Come to think of it, President Erdoğan had a point in clamping down on any dissent. 'At this moment of national significance, there should be unity here in Westminster, but instead there is division,' May said. She had changed her mind over Brexit when she had spotted the opportunity to become prime minister and she couldn't for the life of her understand why other people couldn't be so flexible with their principles.

'The country is coming together,' she continued, waving away the inconvenient truth that no one could remember a time when it had been more split. 'But Westminster is not.' Labour MPs had said they might vote against a deal with the EU if they thought it wasn't good enough. How very dare they!

The Lib Dems – all nine of them – had threatened to grind government business to a standstill. The SNP had promised to be the SNP. Life had become just impossible for her. Her opponents had tried to take advantage of her small majority, so now she was going to punish them by wiping them out completely.

At this point Theresa almost imagined herself to be a latter-day Winston Churchill. Only her enemies weren't the Hun lining up to push the British Tommies into the Channel at Dunkirk, they were the enemy within. Those MPs who had dared to raise concerns on behalf of the 48% of the country who had voted to remain in the EU would be ruthlessly crushed.

'Our opponents believe our resolve will weaken and that they can force us to change course,' she said, unaware of how sinister she sounded. And looked. Her eyes were almost as dead as her delivery: only by disconnecting from herself could she accommodate the cynicism of her position. 'They are wrong. They underestimate our determination to get the job done and I am not prepared to let them endanger the security of millions of working people across the country.' Quite right. If anyone was going to endanger the security of millions of working people, it would be her and her alone.

I. I. I. The longer Theresa went on, the more the statement became all about her. Her leadership. Her party. Her ego. Towards the end she made passing reference to the fact she had only last month declared she wouldn't be calling a snap general election. That had turned out to be just a resolution she had made for Lent. She had tried and tried to resist the temptation of capitalising on the desperate state of the Labour party, taking the opportunity to force through a hard Brexit that almost no one in the country had voted for and guaranteeing a Conservative government for the conceivable future.

But when push had come to shove, the spirit had been willing but the flesh was weak. In what was left of her heart, she knew that no one in the country really wanted another election and that this was being played out for her own vanity and insecurity, but she just couldn't help herself. 'Politics isn't a game,' she concluded severely. But it was and it is. Her actions spoke far louder than her words.

May convinces MPs that Brexit requires her strong and ignorant leadership

19 APRIL 2017

'To Brexit and beyond,' announced a monotone Theresa May, the hollow sockets that pass for eyes giving nothing away. It didn't have quite the ring of 'to infinity and beyond', but then the prime minister doesn't have the charisma of Buzz Lightyear. Theresa can only aspire to the empathy levels of a cartoon character. She attempted a grin to try to prove she can do emotion but her mouth merely assumed a death-like rictus.

Theresa began the formalities of the parliamentary debate to approve her call for a general election where she had left off the day before. How dare anyone suggest she couldn't be trusted to keep her word? When she had said she wasn't going to call an election what she had really meant was that she wasn't going to call an election until she changed her mind. No one could reasonably expect her to be any clearer than that.

The country was more united than it had ever been, she insisted. That was, united as in united in its divisions. And it was high time that Westminster fell into line. All this democratic dissent was getting on her nerves. War is peace. Freedom is slavery. Ignorance is strength. What this country needed to see it through Brexit was a strong and ignorant leadership. And because she didn't as yet have a clue what kind of deal she was going to be able to negotiate with the EU, no one was better equipped to provide that strong and ignorant leadership than her.

'Now is the time for the country to decide,' Theresa concluded. Several SNP MPs interrupted to enquire why now was also not the time for a second Scottish referendum. The prime minister just stared back blankly. Because the Scots couldn't be trusted to vote the right way. Wasn't that obvious? What was the point of having an election in which you didn't already know the result?

Jeremy Corbyn breathed in deeply, channelling his inner Gabriel García Márquez. 'We welcome the opportunity to have an early general election so we can give the country the Labour government it deserves,' he said, going on to insist that the election would be fought on Tory cuts to the NHS and schools rather than Brexit and his leadership skills.

At times like this, magical realism comes into its own. Though not enough to impress his own backbenchers. Only a couple had bothered to cheer him when he had stood up to speak, and both of them looked as though they had regretted that support by the time Corbyn had finished his opening sentence.

The only moment of real Labour party unity had come earlier in the day when Yvette Cooper had deftly skewered Theresa at prime minister's questions by pointing out that by effectively lying about the reasons for holding an election she had made it impossible for anyone to believe a word she said in the future. Theresa had tried to respond but had given up when she realised no one would believe her.

Corbyn's willingness to vote for a general election he seemed almost certain to lose, rather than give the prime minister the headache of a no confidence vote, prompted the Conservative Desmond Swayne to conclude: 'Turkeys really do vote for Christmas.' It was hard to fault the logic.

Though Swayne shared his leader's outrage that the views of the 48% who voted to remain in the EU should be represented in parliament, he did have one gripe. This was the second time he had taken Theresa at her word and made the mistake of assuring his constituents in the local paper that something wouldn't happen that was now happening. Did she have anything else she wanted to tell him before he went for the hat-trick?

As it happened, she might. After accusing the prime minister of gaming the system for her own partisan, political advantage rather than in the national interest – Theresa hastily crossed herself several times – the SNP's Angus Robertson wanted to know why she was so doggedly avoiding a televised debate. 'I have some breaking news,' he declared. 'ITV have announced they are going to have a debate regardless.'

Could the prime minister confirm whether she was still not planning to take part?

Theresa giggled awkwardly. Part of her longed to say the real reason she didn't want a televised debate was because she was worried she would be so brilliant that she would end up with too big a majority. There again, if she was to do a reverse ferret and give in to the broadcasters, then no one would hold it against her. Changing her mind was what she did.

* * *

A week into the general election campaign, the polls were still predicting the Tories would win a landslide majority. This prompted Theresa May and her team to believe they could conduct the election campaign entirely on their own terms. This included the prime minister insisting that she would not be taking part in any of the televised debates that had been a feature of the past two general elections. The reason that was given for Theresa May's no shows was that they were an unwanted distraction and that what she really wanted to concentrate on was going round the country having real conversations with real people.

This might have been more believable if the prime minister had had a track record of getting out and meeting people. Her actual style was rather more akin to that of North Korea's supreme leader – hence my giving her the nickname Kim Jong-May. She would be driven in to closely guarded locations to address a few local party activists for five or ten minutes in a series of meaningless soundbites. It was the ultimate in cynical, echo chamber politics. Not least because the TV cameras would be carefully placed to make a community hall that was three-quarters empty appear as if it was full.

A case in point had been her visit to Bridgend – a Labour heartland – the previous week. The Tory drill sergeant had been arranging his troops. 'Present arms,' he shouted. Several dozen activists held up their Strong and Stable Leadership placards. After about five minutes, almost everyone had had enough. One bloke near the back complained that his arm was aching. The drill sergeant wasn't happy. He had told everyone to keep their placards up until after the Supreme Leader had left and that heads would roll if her reception wasn't anything other than ecstatic. Starting with his.

Eight minutes earlier than planned, a commotion at the main entrance put everyone on full alert. Once again the placards, all of them identical, were thrust into the air and this time they stayed there as Kim Jong-May was greeted with rapture. She smiled awkwardly. The Supreme Leader isn't entirely comfortable meeting ordinary people, even when they have been hand-picked for their devotion.

'This is the most important election in my lifetime,' she began. Primarily because it was the only one in which she had ever stood as Supreme Leader. Kim Jong-May told herself to relax and try harder to engage with her people, but she wasn't entirely sure how to do so. It was so hard to do empathy when everyone in the room was weak and unstable. She willed her eyes to convey warmth, but they remained ice-cold. 'What this country needs is strong and stable leadership,' she continued. 'And only I can provide that strong and stable leadership.' Anything less was unthinkable.

A few minutes in, Kim Jong-May began to switch off. She had said all she had come to say and really wanted to go home, but she understood there were niceties to be observed so she went through the motions. 'This country needs strong and stable leadership,' she said again, her voice now stuck in a metronomic loop. 'And I am what strong and stable leadership looks like. People say the country is divided, but everywhere I go I see a unity of purpose.' It helps if you only go to places where you are assured of a warm reception.

Strong and stable leadership. Every sentence began and ended with strong and stable leadership. That's all the country needed. Other than a plan. 'We need to have a plan,' she confided. 'And that's why we have a plan.' Though she wasn't prepared to reveal what that plan was. Only that the plan was to have strong and stable leadership. With strong and stable leadership, Brexit, the economy and cuts to services would look after themselves. Because when you had strong and stable leadership it invariably turned out that your plan was the right one even when it was the wrong one. Just under 10 minutes after she had started, the Supreme Leader drew to a close. Five minutes later she had left the building.

Meanwhile, a topic that was unexpectedly dominating the airwaves was gay sex, after the Lib Dem leader Tim Farron had dithered for five days before saying he didn't think it was a sin. Thereafter it became inevitable that every party leader would be asked to clarify their views on the topic.

* * *

Kim Jong-May awkward and incredulous as journalist asks question

30 APRIL 2017

Kim Jong-May clutched her left arm tightly. She was out of her comfort zone. Surely the whole point of being the Supreme Leader was not having to go on television to answer rude questions. Still, too late to back out now. She smiled awkwardly. It was always good to try to appear friendly towards one's subjects.

'Don't the voters deserve better than to be spoken to in soundbites?' asked Andrew Marr.

Don't be silly. What this country needed above all were strong and stable soundbites. 'I believe it is in the national interest to have a strong and stable leadership because only a strong and stable leadership can deliver a strong and stable economy.'

Marr reached for the pistol. Him or her? This wasn't the interview he had been hoping for. It was the one he had feared. 'That does sound rather robotic,' he observed. The Supreme Leader began to relax. Robotic was good. Robotic was strong and stable.

With the Maybot fully activated, the Supreme Leader went on to insist that she wanted nothing more than a country which worked for everyone and not just the privileged few. She'd said that hundreds of times before so it must be true. What about the nurses? Marr asked.

They were poorer than they had been for years and many of them were going to foodbanks.

'There are complex reasons why people go to foodbanks,' the Supreme Leader said tetchily. And what people had to remember was that many nurses were just plain greedy and chose to scrounge off foodbanks when they had spent all their money on super-sized meals at McDonald's.

Sensing she might be straying slightly off message, Kim Jong-May returned to her default settings. Strong and stable leadership. Strong economy. Strength through being strong. Security through being secure. No, she didn't feel it would be a failure if inequality rose under her Supreme Leadership. And yes, she did want to reduce taxes, but the best way of ensuring she could do that would be to give herself the leeway to increase some of them. The power of dialectics. Stability through fragility. Integrity through deceit.

The Supreme Leader fidgeted and looked around anxiously, willing the interview to end. 'Jean-Claude Juncker is reported as saying that you are in a different galaxy in the Brexit negotiations,' Marr remarked.

That was an insult too far for Kim Jong-May. Just because she was on another planet, it didn't mean she was from another galaxy. She was very proud to be the Supreme Leader of the planet Zog. And what the people of planet Zog needed was strong and stable leadership, which is why they needed to give her a mandate to strengthen her hand in the Brexit negotiations. Once the EU realised how much everyone in the UK hated it, Brussels would be bound to give us a brilliant deal.

'Do you think gay sex is a . . .'

The Supreme Leader had been expecting this one and she jumped in before Marr had finished his sentence. 'NO, NO, NO.' She absolutely loved gay sex. Nobody liked gay sex more than she did. Nothing was more strong and stable than gay sex providing it was done strong and stably. To be on the safe side, she crossed herself. She could work out later whether it was more of a sin to say something wasn't a sin when you thought it might be.

A few minutes later, the Supreme Leader found herself in the ITV studios being asked much the same questions by Robert Peston. Might as well kill two birds with one stone. Strong and stable, stable and strong. Strengthen our economy by strengthening her own position. Read her lips. She wouldn't be raising any specific taxes. Though she might be raising some unspecific taxes which she wasn't prepared to specify.

By now Kim Jong-May was displaying some nervous tics. Her eyes twitched as they darted in different directions and her fists clenched and unclenched. Desperation was kicking in. 'One last question,' said Peston, thoughtfully opting to put the Supreme Leader and the country out of their misery. Why would she not be doing the live TV debates?

'Because I want to get out into the community to meet some ordinary people,' she replied. Why, only the previous day she had been up to a forest in Scotland where she had met this awfully nice woman, Ruth Davidson, along with seven of her closest friends, who had all told her that what this country needed was the strong and stable leadership which only someone as strong and stable as her could deliver.

On the way back to No. 10, the Supreme Leader asked the Even More Supreme Leader if he thought the morning had gone well. Lynton Crosby nodded approvingly. She had been more mediocre than even he had dared hope.

* * *

At times it felt as if Labour was trying to match the prime minister's metronomic mediocrity step-for-step with uninspired interviews and events of their own. As with the Tories, most of the election heavy lifting was left to the leader, with occasional cameos from trusted shadow cabinet ministers such as John McDonnell, Emily Thornberry and Diane Abbott. All the other Labour MPs went back to their constituencies to run their own private campaigns that highlighted their local appeal and distanced themselves from the Labour high command. Vocal support for Jeremy Corbyn was kept to a bare minimum.

Nor did Corbyn do much to convince that he wasn't

the electoral liability the polls suggested. His main idea seemed to be to introduce an extra four bank holidays a year and the more he tried to convince interviewers that he was really fired up for the campaign, the more he sounded like a blissed-out yoga teacher. Time and again he also struggled to make a coherent case for his position on nuclear weapons. Unsurprising really, as he gave every appearance of disagreeing with Labour's own policy that had been agreed at previous party conferences. Suggesting that Britain might maintain its nuclear submarine fleet while failing to provide it with any weapons was a compromise that did little to convince anyone that the Labour leader was a man who could be trusted with the nation's security.

And when Corbyn did come up with the potentially vote-winning policy of creating 10,000 extra police officers it seemed like he had caught his own shadow home secretary on the hop. Left hand, meet right hand. The interview that Diane Abbott gave to Nick Ferrari on LBC Radio to explain this new promise was one of the low points of the entire campaign. She began by claiming that recruiting 10,000 extra police men and women would cost just £300,000.

Ferrari sounded sceptical that a Labour government could get away with paying police officers £30 per year and invited her to have another go. Abbott then arbitrarily upped the number of recruits to 25,000 before suggesting that £80 million should more than cover the costs of the programme. A starting salary of £8,000 was more than enough for anyone.

'Has this been thought through?' asked Ferrari. Yes, of course it had. Just not by Abbott who now went on to propose recruiting 250,000 police officers. 'The figures are that the additional costs in year one when we anticipate recruiting about 250,000 policemen will be £64.3 million,' Abbott replied, before going on to throw out other possible recruitment targets of 2,500 and 250 along with several other cost predictions of £139.1 million and £217 million. Who knew? One of these figures might even have been right.

Abbott later explained that her apparent confusion was caused by her diabetes, but the voters did not appear to be in a particularly forgiving or understanding mood. When the snap general election had been called, there had initially been some speculation that the local elections scheduled to take place on 4 May might have to be postponed but they went ahead as planned.

Local elections taking place midway through a general election cycle usually spells bad news for the party in government, as extravagant promises made in the heat of the moment run aground on political and economic realities. Limiting losses to an acceptable level are generally about as much as any government can hope for in such a scenario, but the polls indicated that the Tories were so far ahead of Labour they were on course to make widespread gains.

The results went pretty much to form, with the Conservatives gaining overall control of 11 councils and winning 558 seats. Labour lost 320 seats and UKIP were all but wiped out. Corbyn tried to console himself that the election could have gone worse and his party lost even more seats, but that was rather clutching at straws. For most people, the local elections merely confirmed what they had hitherto only suspected: that Theresa May and the Tories were on track for a large majority on 8 June.

* * *

Labour's hint of a pulse leaves Theresa May unsated

5 MAY 2017

Shortly after 4 p.m., Kim Jong-May stepped out of her bunker to deliver a three-minute victory speech to a few members of the media in a factory in Brentford, west London. Not that it was a victory. Not by the Supreme Leader's demanding standards.

The annihilation of UKIP had been no more than she expected, but there was still the hint of a pulse in Labour. Winning councils in the Labour heartlands of north-east England along with everywhere else in the country was all well and good, but she wouldn't sleep easy until all opposition was ground into the dirt.

The Supreme Leader tried to keep the anger out of her voice. The results were definitive proof that the Evil EU Empire was manipulating the British electorate. Not only had it prevented her from winning by an even larger margin thanks to a dirty tricks campaign, it had forced enough people to vote Tory to lure Britain into a false security that the Supreme Leader would get a landslide victory on 8 June. Left to their own devices, the voters might then be tempted to think it was safe to come out and vote for Jeremy Corbyn. She alone had the power to outwit Jerry.

'The results are encouraging,' Kim Jong-May said. 'But I am taking nothing for granted.' The local election results were neither here nor there, really. All that mattered was her coronation on 8 June. We must not fall into Jean-Claude Juncker's trap. If the country relaxed for even a second then Corbyn's coalition of chaos would take power, all our expats would be left stranded on the beaches of Dunkirk and the EU flag would be flying over No. 10.

Now was not the time for triumphalism. The country craved strong and stable leadership and she was the only person strong and stable enough to deliver it. Her sword would not sleep in her hand until she had smitten every Brussels bureaucrat. The general election was too close to call. The whole future of the country was at stake. If she closed her eyes, she could almost believe she was telling the truth. Besides, what kind of Supreme Leader only got a majority of 100?

John McDonnell was much happier. It had been a challenging night, the shadow chancellor admitted, but Labour had responded magnificently by not being totally wiped out. When you thought about it properly, Labour could potentially have lost every single seat in the country. So to have retained control of councils such as Cardiff, that they had only held for several decades, was a brilliant achievement. One that could only have been possible under Corbyn's remarkable leadership.

McDonnell also spoke of plots. There had been a widespread conspiracy by the mainstream media to accurately report that most voters thought Corbyn was a bit of a liability. So for Corbyn to have only lost hundreds of council seats was a total triumph. Once the Labour leader had had a chance to get his message across personally – all visitors to the allotment welcome – he would march to glory in June. People were just going to love the real Jeremy once they really got to know him.

As the news for Labour got progressively worse as the day went on, with the Tories even making massive gains in the Labour heartlands of the north-east, Stephen Kinnock dared to suggest the results were disastrous. Diane Abbott was appalled by this. Like McDonnell, she could hardly have imagined the day going any better.

'So how many seats have you lost so far?' asked an ITV reporter.

Abbott thought about getting out her calculator, but then decided against it. This was a sum that even she could deal with. 'About 50,' she announced, confidently.

'Erm . . . You've actually already lost over 125,' the reporter pointed out.

'Oh,' said Diane, not quite sure what the problem with that was. 'When I last looked we had lost about 100.' 50, 100, whatever . . . They were basically much the same number. They both had noughts in them, after all.

On her way out of the ITV studio, Abbott received a text. Corbyn had gone missing in inaction and every other senior Labour MP had retired to a darkened room with a large scotch. Would she mind going on to the BBC as well? Abbott was only too thrilled. There was nothing she liked more than being given another chance to make an idiot of herself in public. 'It's not a matter of leadership,' she told *Daily Politics*. 'Everyone I have spoken to thinks Jeremy is marvellous.' There again, she had only talked to his close relatives.

Supreme Leader produces pure TV Valium on *The One Show*

9 MAY 2017

Behind every successful Supreme Leader there's a very rich successful investment banker. Or, for their joint appearance on *The One Show*, beside her. Philip looked relatively happy to be on the sofa chatting to Matt Baker and Alex Jones. Theresa could hardly have appeared more awkward if she tried.

Baker broke the ice with a gentle loosener. How hard was it being the husband of the Supreme Leader? Here was Philip's moment to reveal all. To say it was a complete misery spending hour after hour with a woman whose only conversation was 'strong and stable'. Instead he chose to remain loyal. 'There's give and take,' he said. 'I get to choose when to put the bins out.' Philip is clearly a man whose sense of humour doesn't get many chances to shine.

'There are boys' jobs and girls' jobs,' simpered Theresa. Immediately we were right back in a 1960s chat show. A world where men were boys and women were girls. She didn't specify what the girls' jobs were. Other than being Supreme Leader.

Jones then asked about why the Supreme Leader had changed her mind on her walking holiday in Wales about holding an election. Theresa couldn't really come out and say, 'What would you have done if you found yourself 20 points ahead in the polls?' so she muttered something about doing the country a favour. 'What was the drive back to London like?' asked Baker. The Supreme Leader gave this some thought. The traffic had been quite light on the A5 all things considered but there had been a bit of a snarl up on a contraflow on the M1 outside Luton.

With the interview dying on its feet and most viewers thinking it a pity Philip wasn't the prime minister, Baker announced he was going to stop talking about politics because he wanted to get to know the Supreme Leader a bit better. 'Will we be leaving Eurovision?' he asked. The Supreme Leader was momentarily blindsided as Lynton Crosby hadn't given her a script for this. 'No,' she said eventually. 'But I'm not sure how many points we will get.' She didn't appear to be aware that it was a longstanding tradition for Britain to get next to none.

After a short break, the Supreme Leader went on to say she had met lots of different people from all walks of life while she was growing up in the vicarage but despite that had set her heart on being a Conservative MP. 'It's been said you wanted to be prime minister from a very young age,' Jones observed.

'I don't recognise that,' the Supreme Leader replied.

'I only heard her saying she wanted to be prime minister when she joined the shadow cabinet,' said Philip, not altogether helpfully. The Supreme Leader shot him a death stare. Revealing she had had her eyes on the top job since 1999 wasn't necessarily the look she was hoping for.

To compensate, the Supreme Leader went full Maybot. 'The country needs strong and stable government. The country needs strong and stable leadership. I came from a strong and stable family. The country needs stability.' A look of quiet desperation crossed Philip's face and several million people at home felt his pain. They really didn't know how he did it.

'We've found some old footage of Philip from the 1980s,' said Baker. Had anyone at *The One Show* bothered to turn up the volume it could have been TV gold as the camera had caught Philip delivering a personal hymn to the European Union. But the moment passed silently and both Theresa and Philip breathed a little easier.

By now, Baker was looking at his watch. Even by the anodyne standards of an early evening magazine show this was desperate. Philip tried to liven things up but the Supreme Leader somehow managed to kill every exchange stone dead. Yes he had thought she was a lovely girl when he first met her. The Supreme Leader had felt much the same. 'Very stable, very stable,' she said.

'This is turning into an episode of Mr and Mrs,' said the helpless Baker. It wasn't. It was even worse. Did the red box ever make it into the bedroom? The Supreme Leader didn't think it had. Though she couldn't rule it out coming into the bedroom at a later stage. As long as it was strong and stable enough.

'I do like ties. And jackets,' said Philip, trying to fill dead air.

'That's all we've got time for,' sobbed a relieved Jones, as Theresa and Philip scuttled away, pleased to have got off relatively unscathed.

Five years previously Baker had made a name for himself by asking David Cameron the killer question,

'How do you sleep at night?' There was no need to ask it tonight. The answer was obvious. By playing back recordings of her TV appearances.

Cries of 'Corbyn, Corbyn' filled the hall. He had waited a lifetime for this

16 MAY 2017

No one could ever say that Jeremy Corbyn would die wondering. The same couldn't be said of many of his shadow cabinet colleagues who followed him on to the platform in the atrium of Bradford University student union for the launch of the Labour manifesto. Some attempted a fixed grin and tried to remember to applaud in the right places; others could barely manage that. The shadow defence secretary, Nia Griffith, didn't even make it to the starting line. For them this was an ordeal to be endured rather than enjoyed.

'Let me introduce you to Britain's next prime minister,' said Sarah Champion, Labour shadow minister for women and equalities, keeping her fingers firmly crossed. Better to travel in hope and all that. The audience had no such doubts. These were the faithful and they roared their adoration. A chant of 'Corbyn, Corbyn' filled the hall and Corbyn let it linger for several minutes before raising his arms for silence. He had waited a lifetime for this.

Opinion was shifting towards Labour, he insisted. Imperceptibly to the naked eye maybe, but shifting nevertheless. 'Our manifesto will be radical and responsible,' he said, 'and I would like to personally thank everyone who has contributed to it.' There were too many to name by name as it had been through so many drafts and he was still none the wiser about who had managed to leak several versions to the press the week before.

No matter. He was determined to proceed as if everything was coming as a total surprise to the audience. He had managed to find a way of raising £48.6 billion in tax revenues and he was going to spend the exact same amount on improving public services and raising people's standards of living. Perfect symmetry. No one could call his plans uncosted because John McDonnell and Diane Abbott had gone through the figures with a calculator several times.

The Labour government would get the extra £48.6 billion by asking the wealthiest individuals and businesses to pay a little bit more. Quite a bit more, come to think of it. But never mind, they could afford it. Especially that Philip Green. Anyone earning less than £80,000 wouldn't have to contribute a penny more.

Corbyn knew many people might think they would one day be earning £80,000 a year and could be a bit put off at the thought of paying a higher rate of income tax. But now was the time to get real. That really wasn't going to happen, was it? Even if Labour raised the 'national living wage' to £10 an hour by 2020, most people were still going to be fairly broke. Only Labour could guarantee that people would be marginally less broke than under the Tories.

Corbyn was fairly hazy on some of the details of his nationalisation proposals, though he had been reassured they wouldn't cost anything and he didn't appear entirely sure if he would lift the freeze on welfare payments, but his plans to abolish tuition fees, build 1 million homes and create four extra bank holidays got loud whoops.

Brexit got the briefest of mentions. Labour was committed to scrapping the Tories' Brexit white paper even though it had gone to great lengths to support it in the last parliament. Though it was possible that its own new Brexit bill might still end up looking much the same as the Tories'. But this was all fine detail. A need-to-know basis.

Much of the speech was delivered in a minor key, almost as if Corbyn was going through the motions and had long since reconciled himself to the inevitable. Come the questions at the end, the faithful tried to pick him up. A question about immigration numbers was loudly booed. A man from the *Morning Star* blamed the strongly biased media for Labour's poor showing in the polls and was rewarded with an ovation. Even a reporter from the *Daily Mirror* copped it for asking if it was possible that Labour's problem was Corbyn and not his policies.

Corbyn tried to pick himself up. 'This is not a personality cult,' he said as dozens of his supporters shouted out his name. He had been elected as Labour leader by a very large number of people and he had felt their love. All it would take was a little time. Time was something very few of the shadow cabinet seemed to have. The moment the event concluded, most beetled off before anyone had a chance to quiz them. For some of them this would be the first and last time the manifesto got a hearing in the entire campaign.

* * *

The Lib Dems bizarrely chose to hold their manifesto launch in the evening in a hip night club in London's East End at a time that guaranteed almost no television coverage.

After a few warm-up acts, Tim Farron took the stage. The Tories may have accused Labour of taking the country back to the 1970s, with Labour countering that the Conservatives were heading back to the 1950s, but he was unapologetic about his desire to go back in time. To this time last year, when Britain was still in the EU. But first he wanted to talk a bit about Malcolm, the man he had had a dust-up with in Kidlington the week before, and let everyone know that the two of them had kissed and made up. Not a gay kiss, though it would have been fine if it had been, as Tim was perfectly OK with that sort of thing these days.

Guardian Soulmates out of the way, Tim got down to business. The Lib Dems had a new radical vision of politics. Far too many people in the past had made the mistake of voting for the party they wanted to be in government. Now it was time to vote for the opposition.

Only the Lib Dems could stand up to the Supreme Leader's hard Brexit. Yes, he accepted the result of the referendum – will of the people, yada yada – but come on. Someone had to stand up for the liberal metropolitan elite Remainers. Besides, did those who voted Leave really know what they were doing? Surely they deserved another chance to come to their senses once they had seen how crap everything was going to be?

After a quick promise to be nice to children, the climate and sheep, Tim was off. Bugger it. He'd been having such a good time, he'd forgotten to mention the manifesto.

* * *

Close your eyes and believe in your strong and stable Supreme Leader

18 MAY 2017

A woman dressed as a Dalek with a Theresa May face mask joined the protest outside the converted mill in Halifax as the Supreme Leader's five-car motorcade pulled into the parking lot. She didn't look much like the original. The Maybot has far less personality. And is all the more terrifying because of it. Head down and expressionless, she dashed for a side entrance.

Later than planned because the Conservative battle bus had broken down – someone, somewhere would pay heavily for that – the cabinet filed into the front two rows. The men were identically dressed in white shirts and blue ties and had the rictus smiles of the condemned. Boris Johnson and Philip Hammond took up the rear and sat closest to the aisle. Last in, first out. The campaign slogan of Forward Together, which had been nicked from Margaret Thatcher's 1980 Tory party conference, didn't necessarily apply to them.

David Davis, one of the Supreme Leader's trusted lieutenants, took the stage first to make the introductions. He kept to a simple, approved script of 'strong and stable' before quickly sitting down. Careless talk costs lives at the higher echelons of the Tory party these days. Within seconds the Supreme Leader appeared and the cabinet competed with one another to be the first to their feet.

'Today, I launch my manifesto for Britain's future,' Kim Jong-May began. Not the Conservative's manifesto. Her manifesto. Hers and hers alone. In truth, she couldn't quite work out why she had been even asked to produce a manifesto, but as some of the other parties had already done so she had felt rather obliged to write one herself. Not that it was really a manifesto. She wasn't about to tell people what she would like to happen. She was going to tell them what was going to happen.

The country was facing its gravest crisis since the Battle of Britain and what it needed was a strong and stable leader with a strong and stable plan. She was rather hazy about what that plan might be, because that was on a need-to-know basis. And the country didn't need to know. It just needed to close its eyes and put its trust in the strong and stable leadership of the Supreme Leader. She alone could get the best Brexit deal, even if that deal turned out not to be to have a deal. Because no deal was better than no deal.

'I do not believe in ideology,' she insisted, sounding every bit the deranged Mayist. Her ideology wasn't an ideology because it was the right ideology. She was going to lead the Great Leap Forward of the Great Meritocracy. A country where the most deserving got their proper rewards and the unbelievers were left with nothing. This was her country and she could do what she liked with it. If she felt like intervening in areas of social policy then she would and if she didn't then she wouldn't. The people would hear of her interventions as and when she decided to make them. And not before. She might raise taxes and then again she might not. Wait and see.

There were one or two things she was prepared to put right. Right here, right now. There were far too many immigrants and she was going to reduce their numbers to the tens of thousands. Whatever the cost to the economy. Yes, she knew that the Conservative party had promised and failed to deliver that before, but that had been when she had only been Theresa May, the home secretary. Now that she was Kim Jong-May anything was possible.

It had also been brought to her attention that old people weren't dying in the right kind of way. So she was going to make it much more cost effective for people to croak from a heart attack or cancer than to suffer from dementia. People with dementia had no one to blame for their condition other than themselves and their families should pay accordingly. As for children, she was going to abolish free school lunches and introduce free breakfasts instead. Just because. Children had to learn that Breakfast meant Breakfast.

From time to time, the Supreme Leader glanced towards the supplicant faces in the front two rows. Who were these people? Why were they looking at her so piteously? What did they want, these useless apparatchiks? 'Let us go forward together,' she concluded. Quoting Churchill always went down well even if the man was a bit of a loser compared with her. Queen of all she surveyed. Top of the world, Ma.

Forward together, the Supreme Leader left alone in her five-car motorcade. The cabinet were left to fend for themselves in the broken-down bus.

Dim and Dimmer: two Tory car crashes for the price of one

21 MAY 2017

Damian 'Nice but Dim' Green looked miserable before the interview with Andrew Marr had even begun. Of all the TV studios in all of the world, he had to walk into this one. After spending four weeks hidden away from the public on the orders of the Supreme Leader's High Command, the work and pensions secretary was finally being let out at the very moment when there were signs things were going a bit wrong for the Tories. His one chance to prove to the world that he really did have a fully functioning brain was certain to end in failure.

Marr smelt blood and went for an early kill. Why were there no costings in the Tory manifesto? 'It's a realistic document,' said Dim. It hadn't been the Tories' fault that the Tories had called a snap election and, under the circumstances, this was the best they had been able to come up with in a couple of weeks. The manifesto was clearly a document. It was printed on paper. Therefore it was a realistic document.

'There's an £8 billion hole in your plans for NHS spending,' Marr pointed out. Dim twitched nervously. There wasn't, they were just reallocating £8 billion of existing NHS funding, Dim said. 'That's not true,' said Marr. Dim wisely chose not to contradict this.

Things quickly turned worse when Marr moved on to winter fuel payments. These were completely uncosted as no one knew at what level they would be means tested. 'They aren't uncosted,' Dim said defensively. It was just that they hadn't yet been properly costed. Or rather they might have been but it just wasn't the right moment to let everyone know what the costings were. There was no point bothering voters with loads of numbers just before an important election.

Dim was equally out of his depth when asked to justify the government's social care proposals that targeted those suffering from dementia, rather than adopting those published in the Dilnot report. He struggled to do the sums on what the changes might mean to someone with a £250,000 house in Ashford, rambled about there not being the right financial products in place and implied that we should stop being so negative. Rather than focusing on all those with dementia who would be left with £100,000 – and face it, they would all be too far gone to notice – why didn't we concentrate on all those lucky enough to die of cancer who would be able to pass on £1 million to their relatives tax-free?

'You have been touted as the next chancellor,' said Marr, clearly incredulous at Dim's failure to grasp basic arithmetic. 'Do you think Philip Hammond is doing a good job?' This was one of those questions that didn't require an answer. Compared to Dim, Lurch is an economic colossus.

Why stop at one car crash, when you can have two? Having sent Green on to Marr, the Tories chose to double-down by allowing Boris Johnson on to Peston's show. Dim and Dimmer. If Dim had been hoping to supply some gravitas, Dimmer's tactic was to go for full-on levitas. His hair was even more artfully dishevelled than usual and all that was missing was the clown makeup. He even tried to sneak a quick peek of the questions while Peston's back was turned. Not for the first time, his comic timing was off.

'Why are you picking on people with dementia?' Peston enquired, quite reasonably. Dimmer burbled on, hoping that some vaguely plausible answer might come to him. It didn't. The best he could manage was that the government had to pick on someone so it might as well start with the demented.

Peston appeared as startled by this answer as everyone else and asked if Dimmer had been consulted about the contents of the manifesto before it was published. Dimmer was horrified by the suggestion. Why on earth would the Supreme Leader bother to talk to her cabinet about anything? Though he was thrilled that she had promised an extra £350 million per week to the NHS.

'She didn't say anything of the sort,' said Peston, who was now beginning to realise he was dealing with someone in urgent need of psychiatric help.

Dimmer just smirked. When caught telling a blatant lie, smirking is his default response. At which point, a small smattering of self-awareness began to bubble in what passed for Dimmer's consciousness. He strayed off the strong and stable leader message. Time to mention Kim Jong-May.

'What people have to realise is that the election is a choice between Theresa May and Jeremy Corbyn,' he said. Schoolboy error. Dimmer had just reminded everyone of the clear and present danger to the country posed by him also being involved in the Brexit negotiations. Dim and Dimmer, Dumb and Dumber. But which member of the cabinet is Dimmest? Watch this space.

* * *

The Tories hoped that all the bad publicity over social care costs in their manifesto would die down over the weekend. Instead it intensified, with even right-wing newspapers such as the Daily Mail *and the* Daily Telegraph, *who could usually be relied on to back her up, condemning the dementia tax. Theresa May had miscalculated badly. She – or rather Nick Timothy and Fiona Hill – had believed the Tories were so far ahead in the polls, they could get away with putting policies that alienated the party's core support of the older middle classes in the manifesto. With every interview now dominated by the self-inflicted wound of the dementia tax, the Tories were forced into an embarrassing U-turn.*

* * *

Maybot policy reboot ends in an embarrassing interview meltdown

22 MAY 2017

'Nothing has changed. Nothing has changed,' the Supreme Leader snarled, her eyes narrowing into a death stare, her face contorted and her arms spread wide, twitching manically. 'Nothing has changed.'

Everyone at Conservative party's Welsh manifesto launch in Wrexham saw it rather differently. They had distinctly heard her say she would be reversing the Conservative party policy on social care that she had introduced in her English manifesto launch in Halifax the previous Thursday. Making it one of the quickest manifesto U-turns in history.

'Nothing has changed,' the Supreme Leader again insisted, looking increasingly deranged and unstable. Everyone had completely misunderstood her. The manifesto she had written hadn't been the one that was published. When she had said she was going to make everyone pay all but their last £100,000 on their own care, what she had clearly meant was that she was going to cap the amount they would spend on care. And no, she wasn't going to say at what level the cap would be set because the fake news media would be bound to misrepresent her again and besides her manifesto was completely costed apart from the bits that weren't.

By now the Maybot was shaking so badly that one of her arms fell off. Roadie Nick Timothy rushed on stage with a screwdriver to reattach it. 'Nothing has changed. Nothing has changed.' Except that, in the general confusion, her arm had been put on back to front and was now gesturing obscenely to a sign saying 'Strong and Stable, Forward Together' on the wall behind her.

Were there any other bits of her weak and wobbly manifesto she would like to rip up while she was here, someone asked helpfully. It would be so much easier to get the changes in before the ink had completely dried. 'Nothing has changed. Nothing has changed.' The voice was now tinged with panic as well as anger. She hadn't made a U-turn because she was a strong and stable leader. And even if she had, it would have been a strong and stable U-turn. Not that it was.

'Nothing has changed, Nothing has changed.' Why were so many people calling it a dementia tax? Just because it was a tax that targeted people with dementia didn't make it a dementia tax. That was just more fake news. As was the idea that cabinet patsies Damian Green and Boris Johnson had been sacrificed to the Sunday politics shows in a futile effort to defend it. They hadn't really been there at all. They were just holograms created by the fake news media.

'Nothing has changed, Nothing has changed,' she repeated, until sound recordist Fiona Hill put her out of her misery by unplugging her. And nothing had changed. Changing her mind was the Supreme Leader's trademark.

She had done it over Brexit. She had done it over the budget. And she had done it by calling the general election. It was her very inconsistency and indecision that proved she was a strong and stable leader.

Nothing had changed when the Supreme Leader came up against the BBC's Grand Inquisitor, Andrew Neil, later in the evening. Why was everything going so badly, he asked. Kim Jong-May had a ready response. It was because she was going out to the country and talking to them. And telling them something different every time.

'Your policies are uncosted and half-baked,' said Neil. The Supreme Leader didn't deny it. No point really. 'You've just done a U-turn on social care.'

'Nothing has changed. Strong and stable,' the Maybot replied, trying desperately to count down the clock. A look of resignation crossed Neil's face. He stifled a half yawn before suggesting that this must be the first time a manifesto promise had been broken before a government was elected and that Jeremy Corbyn now appeared to be rewriting Conservative policy.

The Supreme Leader's mouth opened and shut but no words came out. She wasn't used to being spoken to so bluntly. Eventually she came up with an explanation. The manifesto had never been intended to be seen as a policy document. Rather it was just a set of vague principles. Random words that just happened to have been strung into sentences. The idea that anyone was ever meant to take it seriously was 'Fake Claims'.

Neil moved on to the NHS. Where was she going to find the extra £8 billion for the NHS? At this point, the Maybot was taken over by malware. She shrugged. Down the back of the sofa? And what about the £10 billion for NHS infrastructure? Down the back of another sofa?

'You've broken your promises on reducing the deficit and immigration,' Neil concluded. 'Why on earth should anyone believe a word you say?'

'This election is all about trust,' Kim Jong-May replied in one of the greatest acts of self-harm seen on TV. 'That's why I've called the election.' After promising not to.

By the end of the 30 minutes, all that was left of the Maybot was a puddle on the studio floor. A cleaner came in and mopped up. The slops were sent back to Downing Street in a taxi.

* * *

On the night of 22 May a suicide bomber blew himself up at an Ariana Grande concert in Manchester, killing 22 adults and children and injuring 250 more. The following morning Theresa May gave a statement on the attack. She looked haunted, slightly diminished even, as she walked out of Downing Street to address the media. The previous evening she had been a mere party leader campaigning hard to keep her job; now she was having to act as the voice of the country after the worst terrorist attack on the UK since 2005.

It's no easy job to capture the nation's mood in just a few words, but the prime minister did it well. She began steadily, describing in some detail what was known about the suicide bombing at the concert and expressing her sympathy for the victims and their family and friends.

Her voice cracked a little as she said: 'It is now beyond doubt that the people of Manchester and of this country have fallen victim to a callous terrorist attack, an attack that targeted some of the youngest people in our society with cold calculation.' This was what made this act of terrorism particularly shocking. It hadn't been indiscriminate: there was no small, cold comfort to be found in its random senselessness. Instead, the killer had deliberately gone after the softest of soft targets: young girls on what should have been one of their best nights of the year.

She understood that there were protocols to be observed and that one of them was her standing there in Downing Street saying the sort of things she was saying. She knew that some would look at the optics of her appearance and be thinking, 'Well, she had to say that, didn't she?' But just because it had to be said didn't make it any less true. She meant every word.

Yes, we all had every right to be shocked. Like the rest of us, her first reaction to the news had been 'Not again'. But humanity would prevail. The bastards would never win. Never. There were just too many of us on the right side of history. At times, words come cheap to politicians.

These ones looked like they cost the prime minister dear. And were all the better for it.

As a mark of respect, all parties agreed to suspend campaigning for several days. When it resumed, this brief insight we had been given into the prime minister's more human side was nowhere to be seen.

* * *

'It's very clear': May disappears into a dreamland of her own

29 MAY 2017

Dragging her heels on to the studio floor, the Supreme Leader finally got to the business end of the election. Taking questions from real people. And Jeremy Paxman. Not easy for someone who struggles to talk human. She had rehearsed at a rally in Twickenham earlier in the afternoon when she relaunched her campaign after the 'dementia tax' meltdown, but it hadn't gone well. She had just repeated the phrases 'strong and stable' and 'coalition of chaos' over and over again. Maybot 2.0 had sounded very much like Maybot 1.0.

'Happy birthday, Faisal,' the Supreme Leader said to Sky's Faisal Islam, who was compèring the audience debate. Backstage there was a huge cheer from her team.

She had remembered her instructions on how to sound like an ordinary person. And then promptly forgot them.

Her first question came from Martin, a police officer, who wanted to know what she was doing about police cuts. 'It's very clear,' she said, before disappearing into a dreamland of her own. One where crime was changing and policing was changing and everything was changing apart from the Supreme Leader's inability to give a direct answer.

By the time she stopped assembling words into random sentences, Martin appeared to have been rendered so comatose he couldn't even remember what he had asked. Islam stepped in to help. Police numbers had decreased by 20,000 on her watch. The Supreme Leader's mouth worked itself into a rictus smile. Better that than no smile at all. Just.

That pretty much set the tone for the rest of the audience participation section. People asked her questions about the 'dementia tax', winter fuel payments, schools and the NHS and the Supreme Leader did her best to fill the 22 minutes with the deadest of dead air. She hadn't changed her policy on anything because what was in the manifesto was never intended to be policy. It was just a series of vague talking points. And when sometime after the election she had decided what was best for everyone, she would let the country know.

This did not go down particularly well with some members of the audience. One man commented 'total bollocks' while those who were still awake laughed out loud with existential despair. It was that or kill themselves. The Supreme Leader's smile twisted into a silent scream. She wasn't used to such lèse majesté.

When Jeremy Paxman had interviewed Jeremy Corbyn earlier in the evening, he had looked and behaved like a man hellbent on acting as a parody of himself. He had interrupted the Labour leader at every opportunity and turned what should have been forensic questioning into a TV turnoff. Someone had clearly had a word with him in the break and he did at least make an effort to let the Supreme Leader get a word in edgeways. Not necessarily to the viewers' advantage, as she continued to do her level best to say nothing at all.

'You've basically changed your mind about everything,' he concluded after listing all the U-turns the Maybot had made in the last few years. 'An EU negotiator would conclude that you are a blowhard who collapses at the first sign of gunfire.' It was Paxman's one telling intervention of the entire evening. The Supreme Leader narrowed her eyes into a death stare at the sound of more laughter. She then just went back to saying nothing at length until she could hear the studio manager call time.

'That went well,' the Supreme Leader said on the way home. By which she meant it hadn't been the total disaster she had feared.

Corbyn was also reflecting on a decent night's work. The audience had appreciated his warmth and empathy, they liked the sound of his policies and after days of being asked about the IRA, he had come up with an answer that sounded vaguely plausible. And being harangued by Paxman hadn't been that bad. Other than for everyone at home. Paxman's main complaint seemed to have been that he wasn't giving a very good impression of being a Trot. And as that had been one of his main aims, it was job done.

* * *

With Theresa May remaining true to her word not to take part in any debate that involved interaction with anyone else from another party, television viewers were treated to the faintly surreal experience of watching the home secretary, Amber Rudd, defend the government's record against the leaders of the other six parties. If the idea had been to paper over the cracks of the prime minister's lack of warmth and spontaneity, it wasn't wholly successful after it emerged that Rudd's father had died just a couple of days before. Though Rudd was adamant that the decision to go ahead with the debate as planned was hers and hers alone, it certainly didn't do much to make Theresa May look a more caring politician.

* * *

The Hand is left to do the heavy lifting while Maybot reboots

31 MAY 2017

Speak to the Hand. AKA the home secretary. The Maybot was clear that there was a reason she wasn't taking part in the BBC leaders' debate.

She was above it all and was choosing to take part in her own Supreme Leaders' debate instead. Just her and her very own echo chamber reverberating with deathless soundbites.

It was an explanation that hadn't gone down particularly well at the Bath engineering factory she had visited before settling down to put her feet up to watch Amber Rudd do the heavy lifting.

Having stood through several minutes of the Supreme Leader trying to think of arguments why anyone should vote for her in an alienated stupor, the staff only started applauding when the media asked if the real reason she was ducking the debate was because she was afraid voters would get to see how truly mediocre and uninspiring she really was.

The Supreme Leader tried laughing as her minders had programmed her to do, but the only noise that came out was a rusty croak.

'I'm interested that Corbyn is interested in the number of TV appearances he is making,' she said, a reply that left everyone confused.

The Maybot usually deals either in tautology or non-sequitur, but this time her system had crashed completely and she had managed both at the same time. To no obvious advantage.

When pressed to clarify what she meant by this, the Supreme Leader did a quick reboot. The Labour leader ought to be spending less of his time concentrating on his telly appearances and more on the upcoming Brexit negotiations.

'That's what I'm doing,' she insisted to a wall of TV cameras, momentarily forgetting that it was she who had called the general election 11 days before the Brexit negotiations began. Easily done.

The BBC's Mishal Husain got the debate under way by inviting the five party leaders, along with the SNP's deputy leader, Angus Robertson, and the Hand, to make their opening remarks.

'Who do you want to lead this country on 8 June?' asked the Hand, clearly expecting the answer to be someone who couldn't even be bothered to turn up.

Tim Farron was worried the Supreme Leader might be spending the evening peeking through people's windows. He needn't have been. She was safely tucked up at home working on the Brexit plan that someone had inconveniently interrupted.

Thereafter the debate largely became open pack warfare on the Hand, with every other leader, apart from UKIP's Paul Nuttall from time to time – with friends like these, etc. – stepping in to point out that the Tories actually had a piss-poor record on everything from living standards to immigration and social care.

Even the audience was against her, cheering on Corbyn, Farron, Robertson and Caroline Lucas as they ripped into her. The BBC said the audience had been selected to be representative. If so, expect a Labour landslide next week.

At which point, the Hand morphed into the Handbot. A hand stuck on repeat.

'Jeremy thinks he has a magic money tree,' she said three times, hoping that at least one of the barbs would stick. It didn't. In desperation she again appealed to everyone to vote for the Supreme Leader who wasn't there.

Even so, the Handbot was probably making a better case for the Supreme Leader than the Supreme Leader could have made for herself.

The Handbot's trickiest moment came with the last question on leadership, as all the other participants predictably chose to point out that one of the things that most defines a leader is the willingness to show up in person and defend your policies.

'Part of being a strong leader is having a good team around you,' the Handbot said gamely before giving up the unequal battle. Hell, if she was going to stand in for the Supreme Leader, why shouldn't she get the credit?

'Have you not read my manifesto?' the Handbot

announced imperiously. It sounded very much as if a palace coup had just been declared.

The Supreme Leader is dead. Long live the Supreme Leader.

Up against the Maybot, Corbyn struggles not to be a personality

1 JUNE 2017

Call it the quantity theory of personality. The less warmth and spontaneity the Supreme Leader reveals she has, the more engaging Jeremy Corbyn becomes.

There was a time at the beginning of the election when the Labour leader looked a little lost; candidates deliberately kept his face off their campaign literature and all but the most devoted of his shadow cabinet would only appear on a platform with him under duress. But as, despite numerous software updates, the Maybot has continued to crash and burn, so Corbyn's momentum has grown.

For his big Brexit speech, Corbyn chose to come to Pitsea in Essex, a constituency in which 70% voted to leave the EU and in which Labour came third in the 2015 election behind both the Tories and UKIP. The days of him only campaigning in areas where he needs to shore up the Labour vote are long gone. The message being sent out was unambiguous: anyone who voted Leave had nothing to fear; Brexit was safe in Labour hands. There would be no backtracking.

It wasn't immediately clear just how many of the 200 Labour activists who had gathered in the sweltering heat of the wood-panelled hall – the room must double as a sauna in off-peak hours – of the Pitsea leisure centre were Leavers.

All seemed to be there for one thing and one thing only – to catch a glimpse of Jeremy. The anti-personality personality. Corbyn has always claimed that his political career has never been about him, but when you're up against a black hole in the form of the Maybot, then it's hard not to be a personality. Just being able to stand up, look vaguely human and talk in sentences that mean something is all it takes.

Corbyn, closely followed by his Brexit negotiating team of Keir Starmer, Emily Thornberry and Whispering Barry Gardiner, all got a prolonged standing ovation as they walked on stage. The Labour leader took it pretty much in his stride but Thornberry lapped it up. It may only have been reflected glory but after years in the shadows, she is happy to take any glory on offer.

Even Starmer seemed to be affected by the occasion. Often the shadow Brexit secretary can come across as rather dour, but he managed to throw in a few extra whoops and smiles for free in his warm-up act. For a lawyer used to charging by the minute, that was quite some concession.

Then came Jeremy. To the manner born. 'Let me introduce you to my Brexit team,' he began, throwing his star-dust towards the star-crossed lovers. 'Look at their intelligence and confidence.' And in that moment it was possible to look at Keir, Emily and Whispering Barry and imagine them going head to head with the EU. It was certainly hard to think of them doing a worse job than Boris, David Davis and Liam Fox.

Next came his Brexit plan. It was considerably more detailed than the Supreme Leader's Brexit plan – a plan which she insists only she has yet refuses to divulge. He promised an immediate recognition of the rights of all EU nationals currently living in the UK along with the protection of workers' rights, and maintaining tariff-free access, but that wasn't really the point. What mattered most was that he promised hope.

With the Tories offering nothing but more austerity to pile on top of the seven years the country had already endured, he was offering something better. There was a light at the end of the Brexit tunnel. We could leave the EU with some grace. Not by crashing out under the mantra of no deal being better than a bad deal.

Thornberry couldn't resist trying to grab her moment in the spotlight but struck an off note by saying she was even happy to take stupid questions from reporters. When you're just beginning to get the media on side, suggesting they are dim may be a sign of the confidence Corbyn saw in his team but not the intelligence.

But no one cared too much as this was Jeremy's day, Jeremy's show. He laughed – in answer to a question from Sky's Faisal Islam, he joked that the first thing he would say to Angela Merkel was 'Ich bin ein Corbyn' – and he was entirely relaxed. Campaigning is what he enjoys most and he was loving every minute. Why wouldn't he when everyone in the room wanted a selfie? Nor was he in the mood to talk about possible post-election deals with other parties as he was planning to win outright. That still might seem a little fanciful, but far less so than a few weeks ago.

* * *

On the night of Saturday 3 June there was yet another terrorist attack, this time in London. A van ploughed into pedestrians on London Bridge before coming to a standstill. Three terrorists, armed with knives, then jumped out and started randomly killing people in the vicinity before these were shot dead by police. Eight members of the public were killed and another 48 injured. With the general election less than a week away, campaigning was only halted for one day. The best way to show that Britain's democratic values could not be destroyed by terrorism was to let the democratic process continue as normal.

* * *

Maybot malfunctions under pressure over disappearing police

5 JUNE 2017

The Supreme Leader had never been more clear about anything. The country was talking about one thing and one thing only. Brexit. So she had come to the same library in the Royal United Services Institute in Whitehall where she had launched her leadership campaign almost a year earlier, to talk about Brexit. That's what the public was demanding and that's what the public would get.

There were a few puzzled faces in the audience. They were under the impression that what most people had been talking about over the past couple of days was Saturday night's terrorist attack in London and they had reasonably assumed that the Supreme Leader might have something to say about it. Apparently not. 'More than ever, the country needs strong and stable leadership,' she said. And that was why she was calling on everyone to strengthen her hand so her leadership could be even stronger and more stable. The Maybot was back up and running.

Mistaking the groans of resignation and despair in the room for confirmation that her message of reassurance was getting through, the Supreme Leader went on to deliver much the same non-speech she had repeatedly given over the previous seven weeks. The same sentences that never quite made sense even on their own. Let alone when they were connected to all the others.

She alone had a Brexit plan. A plan she couldn't fully disclose, other than to say no deal was better than a bad deal. Jeremy Corbyn didn't have a plan because his plan was different to hers. 'We will show leadership, because that is what leaders do,' the Maybot concluded, her algorithms no longer fully operational. 'There is no time for learning on the job.' This was the closest she came to saying anything heartfelt. She'd been trying and failing to learn on the job for 12 months.

Only towards the end of her speech did the Supreme Leader make any proper mention of the London attacks. Enough was enough. She had done everything she possibly could to help the police in her six years as home secretary and it was just a pity they weren't a bit more grateful.

Under her strong and stable leadership, Britain had never been more safe. Even if it didn't feel that way. No one cared more about the country's security than she did. She just had her own idiosyncratic way of showing it. Instead of focusing on the terror plots that had succeeded, why couldn't everyone just concentrate a bit more on the ones that had been thwarted? 'I have the vision,' the Maybot said. The blurred vision of an artificial intelligence without the intelligence.

Much to her surprise, none of the questions that followed her speech were about Brexit. They were all about terrorism and security. The Supreme Leader was patience personified as she failed to answer any of them. Had she been wrong to cut police numbers?

Not at all. The relentless focus on numbers was missing the point. What really mattered was that she had given the police extra superpowers. Some were being trained to have x-ray vision. Some were learning to fly with magic capes. Some could literally bi-locate and be in two places at the same time. Some had been equipped with little suckers on their feet that allowed them to walk up the side of buildings. Some had been given special weapons that fired spiderwebs. Some had also been given invisibility cloaks, which was why it was easy to imagine there were 20,000 fewer of them.

An engineer hastily tried to update the Maybot, but only succeeded in making things worse as her faculties became ever more unreliable. Her voice recognition software couldn't recognise the words 'Donald Trump' when she was twice asked if the US president had been wrong to criticise Sadiq Khan for doing a bad job. 'The London mayor is doing a very good job,' she monotoned. But what about Donald Trump? 'The London mayor is doing a good job.'

'What would Donald Trump have to say for you to disagree with him?' one journalist finally asked. The Supreme Leader looked confused. Who was this Donald Trump? 'The London mayor is doing a very good job,' she said yet again. So would someone – the US president, say – who criticised Khan be wrong? 'I suppose so,' she muttered through gritted teeth. The special relationship had never appeared so pathetically and abjectly one-sided. But it was what now passed for strong and stable leadership.

It had come down to this: vote Maybot, she's a bit better than Corbyn

7 JUNE 2017

Things didn't get off to the best of starts. The Supreme Leader had wanted to spend the last day of campaigning proving she was at least half alive, only to be largely ignored by everyone at the Southampton bowls club where she had dropped in for a morning cup of tea. 'It's nothing personal,' the club captain told her. 'It's just that no one actually realised you were here.'

Her next stop, in a windowless shed on a Norwich industrial estate, didn't go a whole lot better. 'I believe in Britain,' she said in her trademark Maybot tone that suggested she believed in nothing whatsoever. Behind her, there wasn't even a flicker of interest on the faces of the few dozen Tory activists who had been press-ganged in as extras. Just a studied resignation.

'There are great things we can do together, you know,' the Maybot continued. There was a pause as she tried to think of what they might be. Nada. Ah well. 'Who has got the plan to get on with the plan?' she continued, clearly hoping someone might remind her that it was meant to be her. They didn't. Instead there was an awkward silence. The Supreme Leader then remembered she had a 12-point plan. Point one was to have a plan. Point two was that point one should be followed by point two. She didn't get to point three.

The Maybot went to her default settings. She doesn't have any greatest hits so she has to settle for night-time, middle-of-the-road filler on a local radio station. 'Coalition of chaos. Jeremy Corbyn.' The Labour leader's name was followed by an elaborate grimace. One that was intended to be endearing and personable but merely made her look as if she was experiencing a terrifying power surge. After one of her software engineers had gone round the crowd with a cattle prod, there were eventually a couple of dry laughs.

'One of the best things about this campaign is getting out and meeting people,' the Supreme Leader concluded. It was just a shame the people couldn't say the same thing about getting out and meeting her.

Luckily there was never going to be any danger of her meeting any real people at her final rally of the campaign, at the conference centre next to Birmingham's National Motorcycle Museum. Outside, the logo read: 'Where legends live on.' It should have added: 'And where dead-beats go to die.' The whole cabinet had been forced to make an appearance, many of whom may find themselves out of a job by Friday. Philip Hammond, Liz Truss, Liam

Fox and Andrea Leadsom all performed gratuitous acts of self-harm by applauding their own imminent demise.

Boris Johnson came on first. Arms waving, hair askew and with a profound sense of relief that he had some purpose in life after all. He may not be much good as foreign secretary but he makes a half-decent warm-up act. 'Do we want Jeremy Corbyn to be the next prime minister?' he said. 'Nooo,' everyone cried. The Tory campaign had come down to this. Vote for the Maybot. She may be rubbish, but she's a bit better than Corbyn. 'Please give a warm welcome for our wonderful prime minister.'

It took a while for the Supreme Leader to realise Boris was talking about her, but eventually she bounced up onto the stage, followed by her husband, Philip. That little run seemed to use up all her battery reserves because she spent the next 15 minutes sucking all remaining energy out of the room. There was no sense of euphoria or inspiration. Just relief that the whole thing was coming to an end. Intermittently she would get arbitrary rounds of applause mid-sentence – no one really knew when they were supposed to clap – as she spoke with the same sense of despair as she had done for months. 'We have the vision, we have the plan and we have the vision,' she concluded. Though not the vision to know she had already had a vision.

With one last call for everyone to vote on Thursday, the Supreme Leader made her final exit. Next time she had to speak in public, she would almost certainly be prime minister once more. Quite some achievement, she thought to herself. It's not every politician who can win a general election by being so mediocre. God stand up for Maybots.

* * *

Over the course of the last two weeks of the election campaign, the polls had shown some signs that the gap between the Conservatives and Labour were narrowing. But nothing to suggest a major upset. Rather than forecasting an overall majority for Theresa May in excess of 100, most pollsters were predicting one of between 60 and 80 seats. Not quite as all-conquering as the prime minister had hoped for when she had called the election in April, but still respectable enough to vindicate her decision to do so.

As is traditional on the day of an election, there was no campaigning allowed. Nor was there any political reporting in the media. So the leaders went back to their constituencies to await the results. It was the lull before the storm.

* * *

The Maybot asked us to strengthen her hand over Brexit – we declined

9 JUNE 2017

As the exit poll was announced on the stroke of 10 p.m., the opinion room at ITN went into a state of shock. Only a few stunned cheers from Labour supporters broke the silence. The script of the general election had been shredded.

If the polls were anywhere near correct, the Supreme Leader had blown a 20-point lead in seven weeks and would end up with fewer seats than David Cameron in 2015. The Maybot had asked the country to strengthen her hand in the Brexit negotiations and the country had replied: 'If it's all the same with you, we don't think we'll bother.'

Once the exit poll had been released, I made a beeline for Camilla Cavendish, who worked for Cameron in the No. 10 policy unit. Surely her old boss must be feeling a little bit of *schadenfreude* at the Maybot's apparent demise? 'Oh no,' she said loyally. 'He's really not that type of person.'

But one member of the *ancien régime* wasn't quite so good at disguising their emotions. George Osborne was an ITV studio guest and there was a definite sparkle in his eye as the Tories were predicted to win 314 seats. But after the momentary sense of elation and vindication, there was a longer expression of barely concealed regret. If he hadn't been quite so quick to abandon his career on the backbenches for the editorship of the *Evening Standard,* he might have been one of the frontrunners to take over the Conservative leadership within a matter of weeks.

By now it was clear that no one really knew what they were talking about. Of all the outcomes that had been rehearsed, the possibility of a hung parliament never really featured. On the Google screen, the main subject trending was: 'What happens next to Theresa May?' It was a question that would go unanswered throughout the night. As would most others. The young people had turned out in large numbers and everything was up for grabs. The hard Brexit the Maybot had considered a done deal was now a distant memory. As were Angus Robertson and Nick Clegg.

Ed Balls, another studio guest, appeared to be experiencing similar mixed emotions to Osborne. Ecstatic at the apparent Labour revival and gutted not to be at the centre of it. He had come prepared to write Jeremy Corbyn's obituary and was having to ad lib tentative praise.

But the longer the night went on, the more the personal disappointment dissipated. Tribal loyalties die hard and watching Tories suffer the hubris brought on by their own complacency was just too much fun. 'It's all a total mess,' Balls said happily.

It was hard to know which party was more caught on the hop, the Tories or Labour. Both sides were lost for words. Michael Gove was first to rally to the Conservative cause by pointing out that exit polls could be wrong and it was too early to rush to judgment. But he still looked as if he knew the game was up.

Even if the Maybot were to squeeze over the line with a narrow majority, her authority would be destroyed. She would be a laughing stock in the country and the Tories would never forgive her. As Gove spoke, Lynton Crosby was quietly handing back his knighthood. No one would ever trust him to run a general election again. A red-eyed Liam Fox could barely remember his name. The studio manager passed him a tissue to wipe his tears and bundled him into a cab.

Shortly after midnight, John McDonnell appeared, trying his best not to gloat. The shadow chancellor has never knowingly let a smile grace his mouth, but there were definite signs of pleasure.

He too was insistent that it was still too soon to make any predictions. But it wasn't quite. The one prediction everyone could safely make was that almost every polling company had got their sums wrong yet again. Labour was cleaning up in London and even winning seats in Scotland.

The only person who seemed to be almost entirely unbothered by the turn of events was Stanley Johnson. 'Isn't this exciting?' he said cheerfully. It was also possible that his son Boris was feeling much the same way.

Speaking at his count at 3 a.m., Boris was at his most un-Boris like. No gags, just his best attempt at statesmanship. Even to the untrained eye it looked like a naked leadership bid. The Conservatives may be having a nightmare, but Boris wasn't going to pass up an opportunity.

Half an hour later, the Maybot turned up for her own count in Maidenhead. She looked straight ahead, trying not to catch anyone's eye. Her speech spoke of stability but indicated the exact opposite. At best the Tories would be hanging on only with the support of the DUP.

The Maybot had achieved what everyone had imagined impossible. She had inflicted more damage on the Tory party than anyone other than Tony Blair. Overnight the Supreme Leader had proved herself to be anything but supreme.

The Maybot is trapped in the first phase of election grief – denial

9 JUNE 2017

No one could say they weren't warned. The Supreme Leader had promised a coalition of chaos if she lost six seats and a coalition of chaos was what the country was getting. What she hadn't made clear was that the coalition of chaos would be all hers.

After a morning's work of emergency repairs to her circuits, which had overloaded the night before, the Maybot was eventually in a fit state to meet the Queen shortly after 12 o'clock. Her husband Philip put her through her final tests. 'Who are you?' he asked.

'I am the Supreme Leader,' the Maybot replied, rather more confidently than she felt.

'What do you want?'

'Strong and stable. Strong and stable'.

There were still a few software adjustments to be made but they would have to wait, as the car had already been parked outside the front door of No. 10 for more than 20 minutes.

The Maybot and Philip walked briskly to the car, refusing to acknowledge any of the reporters penned in on the other side of the road. They didn't even acknowledge each other. There wasn't anything much left to say.

Well under half an hour later – considerably shorter than many people had expected – the Maybot returned from her audience with the Queen. It soon became clear why. Once she had got to the palace she had completely forgotten what it was she had come to say. She was trapped in the first stage of grief. Denial.

There had been no election. She hadn't blown a 20-point lead in the opinion polls in just over seven weeks. She hadn't just run the worst campaign in living memory. She hadn't published a manifesto that had needed to be pulped before the ink was dry. Everything was normal. Nothing had happened. She was still the Supreme Leader. All was well.

The Maybot made her way slowly towards the wooden lectern set up outside the front door of No. 10. Her wheels often found it difficult to cope with uneven surfaces. A helicopter hovering overhead made it difficult to pick out the Supreme Leader's words. No matter. She didn't really have anything of interest to say.

'I will now form a government,' the Maybot murmured in a catatonic monotone. 'A government that can provide certainty and lead Britain forward at this critical time for our country.' Government. Certainty. Forward. Not the three words that were on anyone else's tongue. It was as if she had been awoken from a seven-week cryogenic state and had decided to mix things up just for the hell of it.

Being in government with . . . with . . . with . . . She couldn't remember the names of any of her cabinet colleagues. Probably because she had never taken the trouble to learn them. But whoever she had been in government with certainly didn't warrant a mention. She was the Supreme Leader and they were just nothing. Strong and stable. Strong and stable.

The Maybot dimly remembered something called Brexit. She had called for the country to strengthen her hand in the Brexit negotiations and the country had listened. By telling her to get rid of a couple of dozen of her own MPs – natural attrition, she reassured herself – and replace them with eight members of the Democratic Unionist party. That would show the EU who was boss. Brussels couldn't fail to be impressed by a bunch of anti-gay, anti-abortion climate deniers. Which reminded her. Perhaps she ought to have a chat with the DUP about the new arrangements sometime. Not now, though.

'Over the next five years,' the Maybot continued, oblivious to the fact that many of her colleagues were giving her five days at most. Over the next five years. Or for ever. Whichever was longer. The Supreme Leader would be strong and stable. And Britain would reach the Promised Land.

With her batteries running on empty, the Maybot spluttered to a halt. Philip stepped forward to push her back inside No. 10. Larry the cat was already on the phone to his therapist. Three owners in a year would play havoc with his abandonment issues. The staff at No. 10 burst into a spontaneous round of applause. It was fully deserved; never in British political history had a prime minister so spectacularly misjudged a post-election speech.

* * *

The disastrous election result immediately put Theresa May's long-term future as prime minister in doubt. Much to the enjoyment of Labour supporters and some of her enemies in the Tory party. George Osborne, the former chancellor whom May had sacked on her first day as prime minister and was now editor of the Evening Standard, *could hardly contain himself on* The Andrew Marr Show. *No one could remember the last time he looked quite so happy.*

The election couldn't have turned out any better and

George was determined to enjoy every minute of his chance to smash up the last remaining working components of the Maybot. 'Theresa May is a dead woman walking,' he said, making no effort to conceal his elation. The only question was how long it took for the computer and the Tories to say no. He looked as if he wouldn't mind if she carried on for a little while yet, if only to prolong her suffering.

'Her promise that no deal is better than a bad Brexit deal is now dead in the water, as the DUP will never allow that,' George continued, punching the air in triumph. Come to think of it, there weren't any bits of the manifesto that weren't dead in the water. The few sections that hadn't been rewritten during the election campaign would certainly have to be rewritten now.

There was also the distinct possibility that Theresa May was a dead woman not walking as she chose to lay low and avoid the chance to explain what had gone wrong to the Sunday politics programmes. She left that job to the defence secretary. In any Tory shitstorm, it's a fair bet that Michael Fallon will be sent out to steady the sinking ship. Primarily because Mogadon Mike is usually too dopey to know there is a shitstorm going on. Sure enough, he began by saying how brilliantly the Tories had done to win the most seats and he was looking to Brexit carrying on as normal.

At which point, Marr had to break it to him that the Tories had actually had a disastrous election and that they were going to have to be propped up by a bunch of gay-hating, climate-change denying, religious bigots. Mogadon Mike appeared genuinely astonished by this piece of news and quickly ad libbed that he couldn't possibly comment on any deal because he didn't know if there was one. But assuming there might be one soon it would probably be best for all concerned if the Tories didn't publicly reveal what promises they had made to 'our friends the DUP'. Under the circumstances, friends was probably not the ideal choice of words.

There may have been moments over the weekend when Theresa May thought about jacking the whole thing in and resigning as prime minister. That was never a realistic option. The last thing the Tory party needed was another general election, which the Labour party would probably win. The voters hadn't wanted the last election and they certainly didn't want another. A second general election within a matter of months could only end in a punishment beating for the Tories.

So Theresa May had to stay in office, at least until such a time as the Tories had another leader lined up. And preferably until after Brexit in 2019. Even the most committed Leavers now accepted that Brexit was – in the short term at least – going to prove toxic, so better to leave Theresa in place to take the hit than risk tarnishing anyone else's reputation. All of which meant that the prime minister was going to be forced to eat some very public humble pie. Starting with the meeting of all her backbenchers at a meeting of the 1922 Committee.

* * *

The Maybot is rebooted as strong and humble. Stumble for short

12 JUNE 2017

The corridor outside committee room 14 was almost full by 4.15 p.m. With journalists. It was 20 or so minutes later that the first Tory MPs and peers started to arrive for the meeting of the 1922 Committee. Among the first to arrive were Anna Soubry and Nicky Morgan, eager to get a front-row seat for the Maybot's humiliation. Morgan looked particularly bright-eyed and chipper.

Boris Johnson was the first cabinet member to arrive. Almost as if he had a point to prove. For the last couple of days he had been seen out and about wearing London Olympics clobber. Given that 2012 was the last time anyone in the country had found the foreign secretary particularly interesting, this had seemed suspiciously like a stage-managed leadership bid. But for this meeting Boris was 100% behind the Supreme Leader. Or as close as he could get to it.

As was Michael Gove, who was the next cabinet minister to make his entrance. Now wasn't the time to stab anyone in the back. That could wait for an hour or two. For now the newly appointed environment minister was only too happy to guarantee EU subsidies for any farmer willing to grow fields of wheat for the Maybot to run through. Philip Hammond arrived grim-faced and head down. Being allowed to stay on as chancellor only because the Supreme Leader was too weak to sack him wasn't great for his self-esteem.

Three minutes after the scheduled start of 5 p.m. the Maybot arrived, flanked by Gavin Barwell, her new chief of staff and former MP who had lost his Croydon Central seat in the election. She looked grim, and her mood wasn't improved by the less than enthusiastic welcome she received. From outside the room it sounded as if the faint banging was MPs smashing their own heads against the desks.

Several minders stood outside the three entrances to make sure no reporters tried to listen in, but it wasn't long before the first MP came out to update everyone. 'There's no sign of the Maybot,' he said, sounding genuinely astonished. He hadn't realised that the Supreme Leader could be reprogrammed to sound almost human. The software update had come too late for the election, but just in time to save her job. Still, at least she had her priorities right.

The backbencher went on to hold court for a good five

minutes. Was she sorry? Yes, very sorry. Did she actually say she was sorry? He seemed less clear on that. 'She was contrite,' he said, choosing his words carefully. 'She said: "I've got us into this mess and I'm going to get us out of it", and that she would serve the party as long as the party wanted her.' And how long did the party want her to serve? He shrugged and walked off.

Ten minutes later a second backbencher walked out. He looked less than overwhelmed by the Maybot's apologies, but managed to stay more or less on message. Everything was hunky dory. Never better. There wasn't going to be a leadership election, he declared firmly. But if there was, would he vote for her? He didn't answer that. Onwards and sideways. Gove came out tight-lipped, but giving a thumbs up sign. With Mikey that could have meant anything.

Some overly loud cheers from the remaining sweaty MPs crammed into the overheated room marked the end of the meeting. The smiles also looked to be just too wide to be true. Everything was going to be different now. Strong and stable had been replaced by strong and humble. Stumble for short. 'She said she was going to listen and govern,' one MP cooed in a state of near rapture. Such was her sense of wonderment it had momentarily escaped her that it might have been better for all concerned if the Maybot had thought of that before the election instead of after.

Not that the Maybot was too bothered by that. She emerged triumphant, with a wide smile. The Supreme

Leader was still supreme. For the time being. All it had taken was a small slice of humble pie and some gentle reminders. So what if the DUP were a bunch of religious bigots? At least they were the Tories' religious bigots.

And if anyone really wanted to take her job they could try to take it if they thought they were hard enough. But if she went it was going to be Boris or Jeremy Corbyn. It was up to them which they thought was worse for the Tories. The Maybot had looked her party in the eyes and the party had blinked first. One for all and all for the Maybot. Her humiliation was their humiliation.

Maybot's reboot stumbles as PM struggles with self-deprecation

13 JUNE 2017

Sic transit gloria mundi. Only a few of the most loyal Tory backbenchers could bring themselves to raise a lacklustre cheer as the Maybot entered the chamber for the re-election of the Speaker, while the father of the house, Ken Clarke, was greeted with full-throated roars from both sides of the house. Seldom can a prime minister have appeared quite so diminished on a first day back in parliament after a general election.

Even John Bercow couldn't resist a gentle dig as he did his best to appear reluctant to be chosen as Speaker for a third time. He talked of his willingness to serve 'the government of the day'. With rather too much emphasis on the word 'day'. The Maybot's head went down at that. She had been counting on making it to the end of the week at least.

Once Tory MP Cheryl Gillan had completed the Bercow formalities with the obligatory reminder that seven former Speakers had been beheaded, to which everyone had roared: 'More, More' – it's the same gag every time but MPs never seem to tire of it – the Supreme Leader rose to address the nation. She began by congratulating Bercow on his re-election. 'At least someone got a landslide,' she said. Even with 'SELF-DEPRECATING JOKE' clearly marked in capitals in the margins of her speech, she couldn't quite manage to coordinate the words with a genuinely warm smile.

With the newly installed Stumble – Strong and Humble – programme still showing signs of teething problems, the Maybot went back to her default setting of denial. The election had actually turned out pretty well, she suggested, because parliament was now more ethnically diverse than it had ever been in the past. So well done her. That was one in the eye for everyone who was under the impression she had called the election out of naked party political self-interest.

'The country is still divided and some people blame politicians for this,' the Supreme Leader continued, sounding mystified as to why this might be. No one dared point out that this could have something to do with her having spent the past seven weeks making highly personal attacks on her opponents, while promising those who voted for her nothing but more pain and more austerity. It's still early days in the Maybot's intensive grief counselling sessions and there's only so much reality she can take.

She concluded by asking the house to come together 'in the spirit of national unity'. That would be a national unity that puts keeping a Tory government in power above the Northern Ireland peace process. And involves going back on almost everything that had been promised in her manifesto. The Maybot sat down to almost total silence from her backbenchers, most of whom went out of their way to avoid eye contact. One even chose to look at half-naked women playing chess on his mobile rather than look up. Start as you mean to go on.

Jeremy Corbyn was in an altogether better mood. No one has yet told him that he didn't win the election and there was a swagger to the way he ripped into the Maybot. 'Democracy is a wondrous thing,' he observed, before going on to say he hoped the 'coalition of chaos' would eventually manage to come up with a Queen's speech. In the meantime, though, he'd be quite happy to chill out with his mates.

'We look forward to this parliament, however short it might be,' he sniggered. And if everything didn't work out for the Tories, 'Labour is ready with strong and stable leadership in the national interest.' Had this been delivered with slightly more grace it would have been all the more effective. But it was still far too devastating for the Maybot, who was in full Stumble mode and staring blankly at her feet.

She did look up when Nigel Dodds got to have his say. In the past she had never given the DUP's leader in the Commons a second glance, but now she listened in rapture as he spoke of the interesting times ahead in the next five years. Not to mention all the dosh that would now start finding its way into Northern Ireland. Suddenly the Maybot was aware of how clever, how handsome and how statesmanlike Dodds was. How could she not have noticed this before?

* * *

Less than a week after the general election, a fire broke out in Grenfell Tower, a social housing high rise in West London, killing at least 80 people. Both her government and the local council were slow to react to what was a tragically avoidable disaster. Flammable cladding had been used on the outside of the tower block and the government still hadn't got round to implementing the recommendations made by an enquiry into a similar tragedy some years previously.

As some Labour MPs called for corporate manslaughter charges to be brought, Theresa May appeared totally shell-shocked. She failed to visit the scene of the fire the next day and when she did eventually go there she only met members of the emergency services. Victims of the tragedy were understandably outraged to be ignored by the prime minister. Her lack of judgment over the Grenfell Tower fire – when both Jeremy Corbyn and Sadiq Khan had made a point of visiting victims – lent credence to the image of her as a prime minister who was lacking in empathy.

The Maybot could feel her grip on power slipping still further. She hadn't even yet managed to agree a deal with the DUP and she couldn't delay the Queen's Speech a second time . . .

* * *

State opening of parliament a crowning humiliation for Maybot

21 JUNE 2017

The crown wasn't at all happy. Normally it got to sit on the Queen's head; now it was made to ride in a separate limo. The Queen also didn't look best pleased by the lack of pursuivants, heralds and ladies of the bedchamber. Her expression never rose above the miserable throughout. Still, at least she was able to make a statement of sorts by wearing a hat in the style of an EU flag. Suck on that, Maybot. There were even empty seats in the Lords. A threadbare state opening of parliament for a threadbare government.

It was all done and dusted in little more than 20 minutes. The longest part was the wait for Black Rod to summon all the MPs from the Commons. Her majesty looked up briefly to check the body language between the Maybot and Jeremy Corbyn. Not good. Come to think of it, the sexual chemistry between the Maybot and her own party wasn't much better.

The lord chancellor handed the Queen a copy of the speech. Brenda flicked through the largely blank pages with a mixture of distaste and disbelief. Was this all the Maybot could come up with after delaying the state opening by a couple of days. 'My government will . . .' she began, her eyes beginning to close.

It turned out that what her government would be doing most of was dumping large parts of the manifesto on which it had been elected. Out with grammar schools, out with scrapping free school meals and the winter fuel allowance, out with the dementia tax, out with energy price caps. Damn it, the Maybot was even reneging on her promise to reinstate fox hunting. That was the one bit of the speech the Queen had actually been looking forward to.

After limping her way through a series of vague commitments on Brexit along with a promise to unite the country – good luck with that one, she thought, you can't even unite your own party – Brenda hit the home straight. 'My government is committed to . . .'. To scrapping the barrel with a whole load of vague promises on space travel and electric cars that had been made in previous Queen's speeches.

'It's almost enough to make one want to abdicate,' the Queen muttered to Prince Charles, who was standing in for the unwell Duke of Edinburgh, on the way out.

'Great idea, Mummy,' said the Prince of Wales, his ears perking up.

'Only joking. Let's fawk awf to Ascot.'

With the Queen safely at the races, the Commons reconvened two hours later to debate the speech. As is customary, proceedings began with two speeches from backbenchers. As isn't so usual, these were neither sparkling nor witty. Perhaps Tories Richard Benyon and Kwasi Kwarteng had decided it was more appropriate to live down to the occasion and keep things dismal.

Jeremy Corbyn stood up and paused. So many open goals, so little time to score them. He eventually opted to begin on a serious note with the Grenfell Tower fire and the terror attacks, before going on to wonder if it was not a little unusual not to implement any of the key promises in a manifesto. A manifesto that had been deleted from the Conservative website only that morning.

This was a new energised and empowered Corbyn and the Tories didn't quite know how to react to him. For years they had been treating him as a joke; now they were being forced to accept he was a possible future prime minister. They didn't seem to like it much. Corbyn did ramble a bit towards the end, but you can't blame him for getting carried away. The way things are shaping up, there's going to be a lot of days in parliament when he gets the better of the exchanges. Without even needing to be particularly good.

There was a desperation to the roar which greeted the Maybot from the Tory benches. A primal scream of despair. The Maybot only confirmed their fears. After an OK start, she rather fell apart. She didn't seem to know much about Brexit. Or anything else, for that matter. She just went back to her tried-and-tested method of saying nothing of any meaning till everyone tuned out. Maybots are as Maybots do.

When Labour MPs pointed out that the election hadn't actually gone that well for her and she couldn't do a deal with the DUP, never mind 27 EU countries, her memory files crashed. 'I won, I won,' she cried. The expressions on those around her suggested otherwise. Freewheelin' Phil grimaced. Boris yawned. Their time would come.

* * *

Two and a half weeks after the election, the Conservatives finally reached a deal with the DUP. In return for the support of the eight DUP MPs on votes of confidence and financial matters, the Conservatives would give Northern Ireland an extra £1.5 billion in funding over the course of the next parliament. Given that the government had previously insisted there was no spare money for anything, politicians of other parties were quick to condemn the deal. But the Tories were unmoved. £1.5 billion was a small price to pay for ensuring they had the necessary parliamentary majority to stay in power for the next five years.

* * *

Maybot's magic money tree? It'll spread the love in Belfast, says Green

26 JUNE 2017

Damian Green looked at his hands in despair. The first secretary of state had spent the last 20 minutes scrubbing them, but they still weren't clean. He turned to the Maybot, begging her to explain the details of the agreement the Tories had reached with the DUP to the House of Commons. After all, the whole sorry deal had only ever been about keeping her in a job.

The Maybot shook her head. She'd just spent 90 minutes trying and failing to convince the house that she was a fair and serious prime minister, and was out on her feet.

Fair and serious is the new strong and stable.

'This deal is in the national interest,' said Green, hesitantly. The national interest as in the Conservative party's interest. He tried repeating 'national interest' but it didn't sound any more convincing second time round, so he adopted another tack.

Think of it this way. The deal was just a happy coincidence. An alignment of stars. The Tory party had been looking to spend some more money on public services and the Maybot had realised that, though Northern Ireland had already been getting more than its fair share, it was probably due another £1 billion top up just for the hell of it.

And if, as a result of the extra dosh, the DUP kept the Tories in government for a few weeks longer, then everyone was happy. Apart from the Scots, the Welsh and large parts of England.

Labour's Emily Thornberry remained unconvinced that risking the Northern Ireland peace process was a price worth paying and wondered where the government had found the money from. Having accused Labour of having a 'magic money tree' throughout the election campaign, how come the Tories had now managed to find one of their own?

At this point Green began to appear severely out of his depth. The deal was so far above his pay grade, all he could do was to refer everyone on to his superiors. He didn't have a clue if the money was contingent on the Northern

Ireland executive reaching a power-sharing agreement, but even if it wasn't, he was sure the DUP would be more than happy to spread the windfall equally among the republicans because that's the kind of easy-going hippies they were. Peace and love and all that.

As for the extra money, that was easily explained. By maintaining the triple lock on pensions and abandoning plans to scrap the winter fuel allowance, the Tories had freed up more cash for Northern Ireland. The Maybot gently broke it to him that he appeared to have got the wrong end of the stick on the financial implications of those measures and invited him to have another go.

'We can afford this because we have a strong economy,' said an increasingly desperate Green. The government had looked down the back of the sofa and just happened to find £1 billion. And if the Scots and Welsh played their cards right and stopped whingeing then they could expect to be on the right end of a bung the next time the government found it had some spare loose change.

Yvette Cooper and Stella Creasy wanted to know what trade-offs the Tories had made on equalities with the DUP. In particular, what the government was proposing to do about women from Northern Ireland being made to pay for abortions in England.

Green hadn't a clue. This was a matter for the Northern Ireland assembly, he said. It wasn't but he'd been wrong on so much already, one more thing wasn't going to hurt.

A few loyal Tories did their best to defend the deal

with supportive interventions handed to them by the whips. Though Crispin Blunt's suggestion that buying a stay of execution for £1 billion was cheap at the price, and everyone should be popping the champagne corks at the government's negotiating skills, was laced with rather more irony than Green would have liked.

Throughout all this, the five DUP MPs in the house could barely contain their excitement.

'This is a good deal for everyone,' said a grinning Nigel Dodds. Especially the DUP. Green clutched his head. The DUP could at least have tried to look as if they hadn't won the lottery. To rub salt in his wounds, Ian Paisley Jr gently reminded the house that the actual figure the DUP had been promised was £1.5 billion. Sod it, thought Green. The magic money tree could probably run to it.

Corbyn scoffs as Theresa tells tall tales of G20 glory

10 JULY 2017

It's debatable whether any of the other world leaders would have recognised her account of the G20, but as none of them were in the Commons the prime minister was free to make it up as she went along. 'Once again we set the agenda,' she began, to loud laughs from the Labour benches. The Maybot looked hurt. The opposition could at least have given her a chance to say what she had set the agenda in. It had taken a lot of time and effort to organise the seating plan for dinner.

The summit had gone brilliantly. Far better than anyone could have expected. She had been the life and soul of the party and several people had said how much they were looking forward to seeing her again. Though not too soon, they hoped. Everyone had also agreed it would be a great idea if they could do a bit more trade with Britain if everything worked out OK. But if it didn't, no worries.

Her success wasn't greeted with quite the acclaim the Maybot had been expecting. Even her own backbenchers appeared indifferent. After all, when was the last time a world leader had said they weren't interested in furthering trade ties with another country at a G20 summit? The whole point of these things was that very little should happen at them, so everyone could agree a bland communique. As long as no leader badly fell out with another then any G20 could be counted as a success.

Realising she was losing her audience, the Maybot was unable to prevent herself from defaulting to factory settings. 'We are building a global economy that works for everyone . . . As we leave the EU we will be negotiating bold and ambitious trade deals . . .' The familiar mindless slogans slipped off her tongue at random until her batteries ran down.

Jeremy Corbyn expressed surprise that she had found so much to do at the G20, given she had openly admitted she had run out of policy ideas at home and had asked the Labour party for input. Several Tories winced at that. The Maybot's desperate appeal to the opposition to prop up her government hasn't been well received in many Conservative quarters.

The Labour leader then tried to pin her down on some details. Just what sort of trade deals did she think she was going to do with the US? Not least because several countries appeared to have come away from the G20 thinking they were Donald Trump's newest best friend and were top of his birthday invitation list. The Maybot made a note not to answer those particular questions.

While Corbyn was on the subject of Trump, he wondered just how hard she had tried to get him to change his mind about withdrawing from the Paris climate agreement. The Maybot wasn't going to take any more of this. 'We have a strong record on climate change,' she said. Andrea Leadsom nodded her head vigorously, though no one knew why – when she became environment secretary she didn't even know if climate change was real. We never did get to find out what the Maybot had or hadn't said to Trump. Perhaps she had just muttered something under her breath.

The more loyal Tories used the time to read out the scripts handed out by the whips on how brilliant the prime minister had been on modern slavery – it was the first inkling that the G20 had actually spent time discussing this at the weekend. Labour MPs were more interested in trying to ascertain the government's position on Euratom. The Maybot had that covered. We would be both in it and not in it at the same time, and we would be bound by the European Court of Justice provided the EU did away with the word European. Simples.

It was left to Labour's Chris Bryant to ask the one thing everyone wanted to know. How was it that Ivanka Trump had managed to wangle a seat at the G20? Was it true she was a great deal brighter than her dad and had she had work done? Not for the first time, the Maybot missed the gag and chose to take the question seriously. It had been utterly appropriate for Ivanka to take her father's place because the G20 had been discussing her pet subject only that morning. Pet subject as in Take Your Daughter to Work Day.

Bryant sniggered and the Maybot was quick to remind him she had been in Hamburg and he hadn't. What was left unsaid was that she almost certainly wouldn't be at the next G20.

You call that a relaunch? The Maybot's broken record is still not fixed

11 JULY 2017

Imagine that someone had never been daft enough to call a general election. Imagine you're heading a government with a clear majority in parliament. Imagine you're still well ahead in the polls. Imagine a world shaped to your own desires.

The Maybot could. 'A year ago, I stood outside Downing Street for the first time as prime minister, and I set out the defining characteristics of the government I was determined to lead,' she began her speech at the launch of the Taylor report into modern working practices in central London.

'I am convinced that the path that I set out in my first speech outside No. 10, and upon which we have set ourselves as a government, remains the right one.' Nothing had changed.

But everything has changed. The Maybot's operators looked on aghast. This was meant to be her big relaunch. Only it was looking very much like the previous relaunch. Maybot 3.0 was the same as Maybot 2.0, which was the same as Maybot 1.0. She was supposed to be sounding a note of contrition, a willingness to adapt to changed circumstances and yet here she was still insisting she had been right all along. It was just everyone else who had got things wrong.

Her operators tried a quick reboot but only succeeded in getting her to speak in passionless, mindless soundbites. Still it was marginally better for her to say nothing than to say the wrong thing. Besides which, meaningless soundbites weren't an altogether inappropriate response to a report whose main findings were that workers should be renamed 'dependent contractors' and that everyone should be encouraged to believe they could have a good job even if all those in good jobs relied on there being enough people prepared to do the crap jobs to make their lives good. Perhaps some might even think that the Maybot was doing a subtle pastiche of the Taylor report. Then again, perhaps not.

While her system administrators were still busy reprogramming her, the Maybot pressed on. This was a good report. A very good report. Such a good report, in fact, that she was going to take it away and study it very carefully over the summer so that when parliament returned in September she could quietly ignore most of its recommendations. To implement any of the proposals would be to do a disservice to the blandness of the report.

Halfway through her speech, the empathy function briefly kicked in. The Maybot knew what it was like to be on a zero-hours contract. She knew what it was like to be stuck in a job you hated and for which you didn't have the relevant qualifications. She felt the crippling insecurity of knowing you could lose your job at any time. She shared their pain. 'I want people to be able to go as far as their talents will take them,' she sobbed. Or a great deal further in her case.

'We will always be on the side of the hard workers,' she continued, momentarily forgetting that only the previous day she had confirmed that teachers would effectively be getting a pay cut. Still, most teachers were probably a wee bit lazy at heart. Just like the doctors and nurses. Most of them didn't know the meaning of a proper day's work.

Belatedly, the rebooted Maybot flickered and remembered she was supposed to reboot. Now she came to think of it, the election hadn't gone entirely to plan so she would quite like the Labour party's help to get through a government agenda she couldn't get past her own party. Then the memory faded. The election had actually been a great idea because it had got more women into parliament. Mostly Labour MPs. She ended by insisting her government had an unshakeable sense of purpose. Even though everyone in it thought it was on life support.

As the event drew to a close, the BBC's Kamal Ahmed stood up to ask a question. 'Camel, Camel,' the Maybot said absentmindedly. 'I'm sorry, I was thinking of something else.' Some of her operators started openly weeping. The Maybot hadn't even been able to maintain her concentration for the 30 minutes of her own relaunch. Press control-alt-delete.

Maybot's 'little tear' interview: a masterclass in robot ethics

13 JULY 2017

The preparations had gone on all the previous night. Installing an empathy function into the Maybot's operating system had proved a great deal trickier than the administrators had imagined. Time and again, the update had appeared to load satisfactorily only to end with a spinning circle of doom. Finally, just minutes before the Maybot was about to begin her first broadcast interview since the general election, an engineer had found a way to bypass the glitch. It wasn't perfect but it would have to do.

'I'm very pleased to be joining you,' the Maybot told Radio 5 Live's Emma Barnett. She didn't sound particularly pleased, but being pleased was what she had been told the occasion demanded. Being pleased also conformed to Isaac Asimov's second law of robotics. A robot must obey orders given by human beings except where such orders would conflict with the first law.

Barnett did not sound entirely convinced by the Maybot's expression of pleasure, but let it go. When did she realise the election campaign wasn't going to plan? This question threw the Maybot into some confusion, as she still wasn't entirely sure that the election campaign hadn't gone to plan. Yes, it hadn't gone perfectly, but she couldn't put her finger on anything she had actually got wrong. All the incoming data she had received up until the exit poll was published had indicated she was on track for a resounding victory.

'Did you shed a tear at the result?' asked Barnett, going for an early money shot. Proof that the Maybot had some human attributes would make headline news.

Had it been a tear or had she just sprung an oil leak?

The Maybot hesitated. She couldn't quite remember. But never mind. 'Yes, a little tear,' she said. A tearette.

'Was there any moment when you thought about stepping down?' Barnett continued.

The Maybot looked confused. What kind of idiot question was that? Hadn't Barnett studied the third law of robotics? A robot must protect its own existence as long as such protection does not conflict with the first or second laws. Of course she hadn't thought about stepping down. She had won the election and it was her duty to form a government, otherwise she might self-destruct and be scrapped.

Even though the result had not turned out anything like she had been led to expect, she would still do everything the same all over again. Apart from the bits she would change. It hadn't been her fault that some of the messaging of the Conservative campaign had been a bit off, as she hadn't been in charge. She was only the prime minister, after all.

'Aren't you out of touch?' an incredulous Barnett observed. Not at all, the Maybot insisted. She definitely didn't regret calling the election because the Tories had gone on to win Mansfield. No Tory leader had managed that in years. And now she was back in No. 10 she had proved her humility, by showing she had listened to young people's concerns about the lack of housing by putting absolutely nothing about social housing in the Queen's speech.

Although Barnett had teased out some flickering signs of personality, she still had doubts about the Maybot's humanity. She gave it one last go by asking if she was a feminist.

'Er . . . I have said that before,' the Maybot replied unsteadily, as her circuits began to overload and fail. She was soon back on restored factory settings. Lifeless, deadbeat and only able to talk in the most mindless generalities.

The deal with the DUP was worth it because it gave the country a strong and stable government and, besides, the £1 billion wasn't that much because the economy was growing so fast it appeared to be contracting. Brexit was going to be a success because if you closed your eyes and hoped for something hard enough it invariably came true.

Everything she had ever done had definitely been the right thing to do, because doing the right thing was what she was programmed to do. The first law of robotics made that clear: a robot may not injure a human being or, through inaction, allow a human being to come to harm.

'What would you say now to your 16-year-old self?' Barnett concluded. The Maybot struggled with this. 'Gosh,' she exclaimed. 'Believe in yourself? Do the right thing?' Wrong. The correct answer was 'be careful what you wish for'.